LOST WAX CASTING
of jewelry

Frontispiece. The Gloucester
Candlestick. Courtesy of The
Victoria and Albert Museum.

LOST WAX CASTING
of jewelry

Keith Edwards

HENRY REGNERY COMPANY
CHICAGO

Library of Congress Cataloging in Publication Data

Edwards, Keith.
 Lost wax casting of jewelry.

 Bibliography: p.
 Includes index.
 1. Jewelry making—Amateurs' manuals.   2. Precision
casting.   I. Title.
TT212.E35 1975   739.27'4   74-30115
ISBN 0-8092-8332-8

© Keith Edwards 1974
First published in England in 1974 by Mills & Boon Limited, London
First published in the United States in 1975 by
     Henry Regnery Company
     180 North Michigan Avenue, Chicago, Illinois 60601
Manufactured in the United States of America
Library of Congress Catalog Card Number: 74-30115
International Standard Book Number: 0-8092-8332-8

Contents

Acknowledgements

For far too long has an aura of 'mystery and myth' surrounded this area of jewellery construction, due in no small part to the dearth of published work on the subject. In offering this book as an introduction to contemporary Lost wax casting techniques, I should like to express gratitude to my friends and colleagues, Graham Davies, who first introduced me to the topic and whose jewellery creations have been a constant source of inspiration and Brian Whalley for his unstinting help with much of the photographic illustration.

To the Worshipful Company of Goldsmiths for permission to include illustrations from their very extensive collection.

To Peter Gainsbury, the Technical Development Officer of the Goldsmith's Company, for his generous help and advice.

To Gilian Packard, John Donald and Andrew Grima and others for permission to include illustrations of their work.

To Richard Ogden for his permission to include a selection from the Burlington Arcade 'Ring Rooms'.

Finally to the numerous companies listed in the Buyer's Guide section, without whose generous co-operation this book could not have been completed.

1   Introduction

To describe the creation of jewellery by lost wax or cire-perdue as a new modern craft would seem to be a contradiction in terms, as this form of casting has been traced back over 3,000 years B.C. Yet it is only within the last few decades that the process has become more widely used in this country, and even then, only infrequently outside the Jewellery and Dental trades and a few specialised Colleges of Art.

Basically the process is a simple one. A facsimile (the pattern) of the piece to be cast is first modelled in wax and is then surrounded by a liquid slurry of refractory plaster called investment (see Fig. 1). When set, the investment, usually surrounded in a metal cylinder, is heated to melt out the wax and leave the mould cavity.

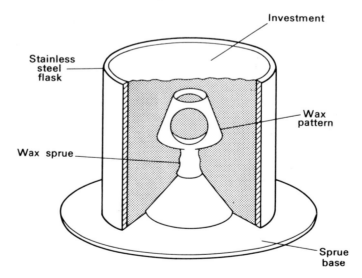

Fig. 1. An invested wax ring pattern.

Subsequent heating will remove residual moisture and eventually oxidise the carbon residues left from the wax. The molten metal is then poured or injected into the mould cavity and will assume the form and texture of the original pattern. After partial cooling, the plaster mould is broken open to reveal the casting.

In his book *Silverwork and Jewellery*, H. Wilson refers to Theophilus Presbyter, a tenth-century monk, whose prolific writings on 'Divers Arts' did much to influence the work of craftsmen in the Middle Ages, and attributes to him the following description for constructing a pair of handles for a silver chalice:

Take wax and form the handles thereof, and model on them dragons, or beasts, or birds, or leaves in whatsoever way thou wishest. On the top of each handle, however, place a little wax, rolled round like a slender candle, as long as a little finger, the upper part being somewhat larger. This is called the 'pour', this thou wilt fix to the handle with a warm tool.

Then take well-beaten clay and cover up each handle separately, so that all the hollows in the modelling may be filled up. When they are dry, again coat evenly all over, and in a like manner a third time. Afterwards, put these moulds near the coals, so that when they get hot thou mayest pour out the wax.

The wax being poured out, place them wholly in the fire, turning the mouth of the moulds by which the wax ran out, downwards. When they glow like coals, then melt the silver, adding to it a little spanish brass.

Taking the moulds out of the fire, stand them firmly up, and pour in the silver at those places where thou pourest out the wax. When they have cooled, break away the clay, and with files and scorpers join them to the chalice.

Although not in the classification of jewellery, one of the finest examples of casting by the cire-perdue process known to the author is to be found in the Victoria and Albert Museum in South Kensington.

Standing just over two feet high, the Gloucester Candlestick (Frontispiece) was cast in the twelfth century from brass or bronze and bears the inscription: Peter Abbot of Gloucester (1104–13).

One can but marvel at the skill of the forgotten craftsman who carved the original wax model, and

8

wonder if, even with all the sophisticated 'hardware' at his disposal, the modern craftsman could emulate such artistry.

In the main, the lost wax method was used for casting statuary and similar artifacts and may be loosely described as a type of shell moulding, the metal being poured into the mould under normal gravitational pressure.

The first use of a liquid investment slurry in 1898 is generally attributed to an American dental surgeon named Philbrook and it is to him, and his successors in the dental profession, that modern jewellery casting practice owes its current popularity and success.

Modern technology and science has produced better waxes, investments and injection methods than those available to the ancient craftsmen, but the principles remain unchanged.

Whereas, formerly, natural waxes such as beeswax softened with turpentine were used to make the original pattern, the modern craftsman uses a range of synthetic and natural waxes remaining workable over a fairly wide range of workroom temperatures.

Modern investment materials come packed in airtight containers with carefully specified instructions for mixing, setting and burnout.

Until quite recently the craftsman relied upon an incredible range of natural substances to invest, or coat, the wax model. Some workers applied successive coats of a slurry of finely-ground clay, loam, sand and pumice, or bath brick. Others used sawdust, cowdung, rotten rag, chopped straw or shredded paper in a medium of plaster of paris.

For large castings, gravity injection is still used as in the past. The method requires considerable skill and experience in positioning the sprues and air vents to avoid trapping air in the hollows of the mould, but for smaller and more intricate work, vacuum, centrifugal and pressure injection methods are now widely used.

Lost wax casting has both an ancient tradition and a modern relevance. It is versatile enough for use on any scale with base and precious metals.

Given basic facilities, such as working surfaces and heat sources, capital expenditure need be minimal. Working materials, such as waxes and investments, are by no means prohibitive in cost, whilst many of the modelling tools may be fashioned from odds and ends.

For a workshop already using normal constructional methods for fashioning jewellery, lost wax casting represents a natural extension in this technique, allowing clean scrap to be used, rather than sold back to the bullion dealer at a loss, at the same time extending the designer's scope into exciting three-dimensional solid forms.

2

An illustrated outline of the fashioning of a silver ring by lost wax or investment casting

For the benefit of readers for whom lost wax casting is a new process, this chapter is devoted to a pictorial description of one method of making a simple textured ring in sterling silver, using both air pressure and centrifugal methods of casting.

The value of the metal used in the finished ring, depending upon its dimensions, would be quite modest.

More detailed description of pattern making, investing, burnout cycles and methods of casting will be discussed in subsequent chapters.

In order to minimise the time and materials required to get the ring to its finished size, one should first ascertain the size of the finger the ring is to fit. The finger is sized using a set of standard ring gauges (Fig. 2), bearing in mind that it is usual to add at least half a size if the ring is to be a fairly broad one.

Let us assume the finger size to be N, then by reference to the table of ring sizes (p. 178) the size, allowing for the broad nature of the ring, would be 17.38 mm.

Making a ring mandrel

A piece of bright mild steel, or aluminium, of about 20 mm dia. and in the region of 100 mm long is machined to the dimensions indicated (Fig. 3). The length is purely arbitrary, but it is wise to standardise on one length, which will make racking and storage more convenient, as a collection of sizes is built up over a period of time.

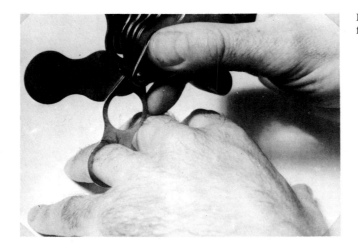

Fig. 2. Checking the size of a finger with a ring gauge.

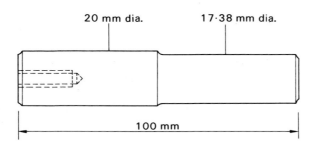

20 mm dia. 17·38 mm dia.

100 mm

Fig. 3. The ring mandrel.

Fig. 4. A ring mandrel with supporting stand.

Time will be saved later if the mandrel is stamped with its size on the reduced end. This is best done immediately the mandrel is machined to size, avoiding later tedious checking of several mandrels with a micrometer until the right one is located.

The threaded hole in the mandrel will enable it to be mounted on a stand (Fig. 4), leaving both hands free to work. For the benefit of those readers who do not have access to a metal turning lathe, both the stand and a set of mandrels may be purchased.

Making the pattern
To make it easier to remove the finished wax pattern, the mandrel is coated with a thin film of light lubricating oil. It is essential that both mandrel and oil are clean before starting, as any particles of grit or dirt will be included in the wax and could result in pitting of the final casting.

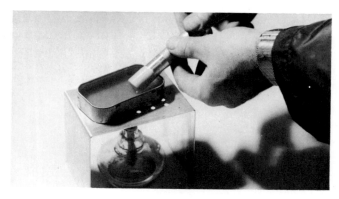

Fig. 5. Building up a coating of wax.

In Fig. 5 the end of the mandrel is being revolved in a container of melted wax until the desired thickness is built up. It should be removed frequently, still being rotated, to allow the wax to set partially as immersion for too long a period will take one back to square one!

Care should be taken not to overheat the wax when using a flame for heating. It should be just liquid and not under any circumstances be allowed to boil.

A safer method would be to heat the container of wax in a water bath, and the use of a thermometer is recommended to find the optimum working temperature of the wax being used.

Fig. 6. The wax coating ready for turning.

Fig. 6 shows the wax build-up completed and time should now be allowed for the wax to harden. Alternatively, the coated mandrel may be immersed in cold water for a few minutes, though this could set up stresses in the wax, causing fracture of intricate detail with certain styles of pattern.

In Fig. 7 the mandrel is shown being gripped in the chuck of a lathe whilst the wax is cut away to a

Fig. 7. Cutting the wax to the required thickness.

13

uniform thickness. The wax is then under-cut until it becomes translucent and the width of the pattern can be readily seen (Fig. 8). The under-cut should not be taken through to the metal or scoring of the metal will be inevitable. The remaining thin layer of wax will also serve to prevent the pattern from slipping during subsequent operations.

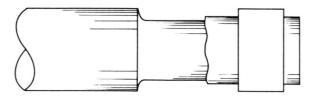

Fig. 8. The wax is undercut to form the width of the ring.

In Fig. 9 the shape is being formed with a scraping tool. A lathe is, of course, very useful for this operation but is by no means essential. One can arrive at the same shape by careful use of wax knife and assorted files, though at the expense of time.

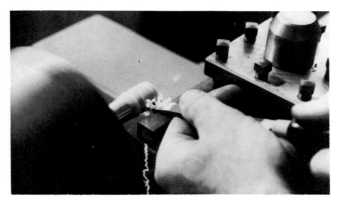

Fig. 9. Forming the wax into a shallow 'D' section.

When the shaping has been completed, the mandrel is screwed on to its stand and the texture scored with a modelling tool (Fig. 10). The spirit lamp shown in the background is used occasionally to melt the wax build-up on the modelling tool, the wax being wiped away with a soft cloth reserved for this purpose.

Good background lighting is useful during this stage, to ensure that no areas of the pattern are left untextured.

Using a warmed knife, the surplus wax is trimmed away carefully (Fig. 11). Should the pattern not slide

14

off the mandrel easily, the shank of the mandrel may be immersed in hot water until the pattern slides free. Extreme care should be exercised during this stage as the wax will be softened by the heat and easily distorted.

Fig. 10. Applying the texture with a wax carver.

Fig. 11. Cutting away the surplus wax with a wax knife.

When the wax has stiffened up again, the pattern may be scrubbed gently with cold water and detergent, using a soft brush to remove small flakes of wax and any dirt that may have been picked up.

Mounting the pattern

Depending upon which method of casting is to be used, a sprue former (Fig. 12) of appropriate size is selected. The former must be cleaned of all old investment. In both methods of casting a single ring it is advisable to insert a metal sprue wire of about 1.6 mm dia. into the hole at the top of the former (Fig. 13). This will give a firm support to the wax sprues and, in the case of pressure casting, will be the diameter of the hole through which the molten metal will be forced. The metal sprue wire is fixed with a blob of sticky wax on the underside of the dome.

Fig. 12. Sprue formers for pressure and centrifuge casting.

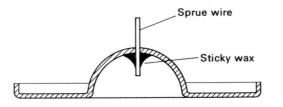

Fig. 13. Fixing the sprue wire with sticky wax.

Where pressure casting is to be used, a small wax ball of about 4 mm dia. is built up by touching a warmed modelling tool on some cold wax and depositing the

16

melted wax on to the sprue wire (Fig. 14). This ball
will eventually form a reservoir cavity in the mould,
which will feed the sprues during casting.

Fig. 14. Building up the sprue
reservoir.

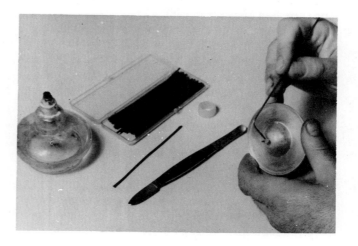

Fig. 15 shows the position of the ball in relation to
the dome of the crucible former. This distance is
critical.

Fig. 15. The position of the sprue
reservoir is critical.

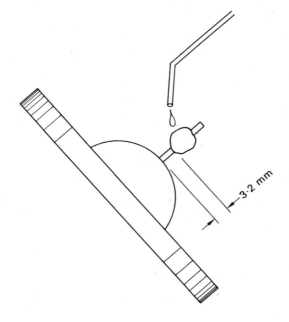

3·2 mm

No such reservoir is needed on the former used for centrifuge casting, but the sprue wire and the top portion of the cone are coated with wax to form a firm base for fixing the wax sprue wires.

Three or four wax sprue wires are now cut to length with a warm knife and fixed on the edge of the pattern so that they slope to form a conical shape (Fig. 16). Sticky wax is applied with a warm tool and the sprues positioned before the wax has set.

Ordinary modelling wax is then touched into the corners to form fillets. These fillets will serve two purposes: first, to strengthen the joint and secondly, to avoid sharp corners being left in the mould when the wax has been melted out.

At this stage it is wise to calculate the amount of metal required to make the casting. Some craftsmen will make an estimate based upon previous experience, but for the beginner it is best to calculate the amount to within the region of say 10 grammes.

Over-estimation of the amount of metal required as a means of 'playing safe' will raise the initial cost which, in the case of gold, can be quite high. Too little, on the other hand, can prove disastrous, with an incompletely filled mould cavity.

A simple method of estimation is to partially fill a graduated measuring cylinder with a known amount of cold water. The pattern and sprues are immersed in the water and the subsequent raising of the level noted. It will probably be necessary to fasten a piece of fine stiff wire to the pattern as most waxes float.

The pattern is removed and pieces of the casting metal added to the water until the water reaches the same level.

A more accurate method is to weigh the pattern and sprues on an accurate balance and then multiply the result by the specific gravity of the casting metal which, in the case of silver, is 10.30.

With both methods, additional wax to form the residual sprue button should be included in the calculation. About 20% of the pattern and sprue weight should be ample.

Fig. 17 shows the pattern being fixed to the sprue reservoir (pressure casting).

Fig. 18 shows the pattern mounted for centrifuge casting. The parts of the former which will come in

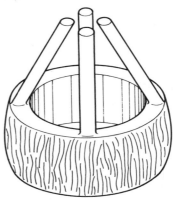

Fig. 16. Four sprue wires are firmly welded to the edge of the ring.

Fig. 17. Welding the sprue wires to the reservoir.

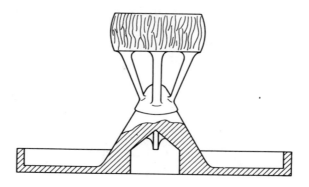

Fig. 18. The pattern mounted for centrifuge casting.

contact with investment should now be coated with a thin film of light oil to prevent adhesion when the former is separated from the casting flask.

Fig. 19. Coating the pattern with debubbliser.

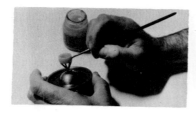

A film of debubblising agent is now applied with a fine brush (Fig. 19). The purpose of this agent is to release the surface tension of the wax, allowing the liquid investment to flow into every detail of the pattern.

The model is then set aside to dry whilst the materials for investing are prepared.

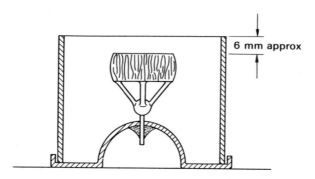

6 mm approx

Fig. 20. Selecting a flask of suitable size.

Fig. 20 illustrates the correct selection of the flask size. Too much depth of investment above the pattern and the gases may not be able to escape when the metal is injected into the mould. Too little and the investment may break away with the force of the injected metal.

A strip of asbestos paper is carefully placed around the inside of the flask (Fig. 21), allowing only a small overlap, and tacked in position with a few blobs of sticky wax.

19

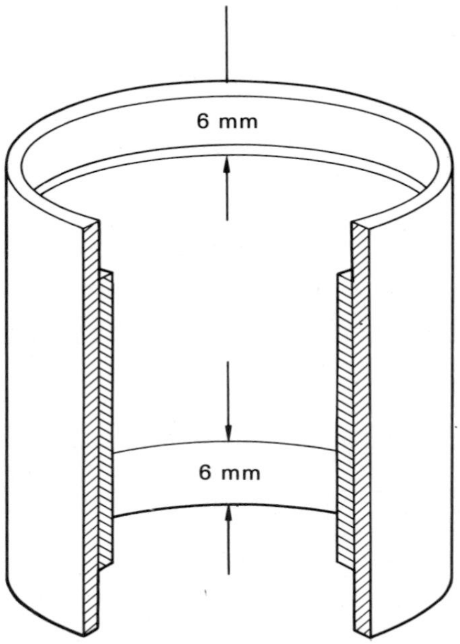

Fig. 21. Lining the flask with asbestos paper.

6 mm

6 mm

A gap of about 6 mm is left unlined at both top and bottom of the flask to act as a key for the investment. The lining will help to maintain the dimensions of the mould cavity through the subsequent stages of expansion and contraction.

In order to prevent too much moisture being absorbed from the investment slurry, the flask is immersed in water and allowed to drain prior to fitting to the former.

A fillet of plasticine will effectively form a seal between flask and former.

Investing the pattern

Fig. 22 shows the basic equipment needed for weighing, measuring and mixing the investment. The ratio of water to powder is fairly critical and manufacturer's instructions should be followed.

The investment is added in 100 gramme amounts to the water, which should be at a temperature of 20–22°C, and is mixed thoroughly with a curved spatula until a uniformly creamy slurry is obtained (Fig. 23).

With the flask resting on a vibrating table, the investment is poured alongside the pattern until it reaches the lowest point of the pattern (Fig. 24).

20

Fig. 22. Basic equipment for weighing and mixing the investment.

Fig. 23. Mixing the slurry with a curved spatula.

At this stage it is wise to vibrate the flask to bring to the surface any air bubbles trapped in the investment. After a short period of vibration the remaining investment is poured into the flask until the level is about 6 mm over the top of the pattern. Another period of vibration will free most of the remaining air.

Should bubbles be left in contact with the wax, they will result in small growths, or 'nodules', on the final casting.

Some 20 minutes later, when the investment has set firm, the surface is roughened to allow easier escape of gases during casting (Fig. 25).

Fig. 24. Pouring the investment into the flask alongside the pattern.

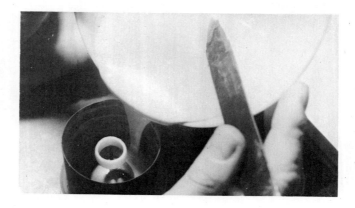

Fig. 25. Roughening the top of the invested flask

Fig. 26 shows the sectional detail of the pattern and flask invested for pressure casting.

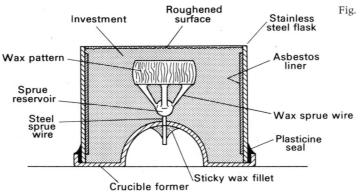

Investment — Roughened surface — Stainless steel flask

Wax pattern — — Asbestos liner

Sprue reservoir —

Steel sprue wire —

— Wax sprue wire

— Plasticine seal

Crucible former — Sticky wax fillet

Fig. 26.

Fig. 27 shows the removal of the metal sprue wire. A gentle flame may be applied to the pin if the former is a metal one. The plasticine seal is now peeled away and the former removed from the flask with a firm but gentle twisting action.

Fig. 27. Removing the heated sprue wire.

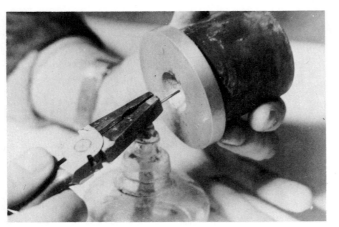

22

Fig. 28a. A hemispherical
depression for pressure casting.

From now on care must be taken to prevent any particles of investment or grit from entering into the sprue hole, as these could be included in the final casting.

Fig. 28 shows the invested flasks for pressure and centrifuge casting. Note the small entry hole in Fig. 28a where the metal is actually melted in the hemispherical depression. The conical depression in Fig. 28b serves to funnel the molten metal into the mould cavity and to act subsequently as a feeding head.

Fig. 28b. A conical depression for centrifuge casting.

Fig. 29. Balancing the centrifuge with the crucible and invested flask in position.

In order that the centrifuge arm will revolve without vibration, the weights are adjusted until a balance is achieved between them and the combined weight of the crucible, casting metal and the invested flask (Fig. 29). For all practical purposes, the weight of the moisture in the investment may be discounted.

The burnout
When the flask has bench-set for about an hour it is inverted, sprue hole down, on a small refractory trivet and placed in the muffle of the burnout furnace, which has been pre-heated to a temperature of about 150°C (Fig. 30). It is left there for at least one hour to dry out much of the free moisture in the investment and to melt out the wax into the trivet.

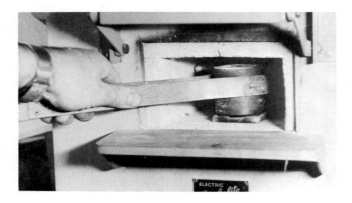

When the wax has been eliminated, the trivet is removed and the flask laid on its side with the sprue hole facing the furnace door (Fig. 31). The temperature should be raised, gradually, to 700°C and held there for about two hours so that the mould temperature is even and the carbon residues left by the wax have been oxidised away. Note the discolouration of the investment in its unfired condition.

Fig. 31. Sprue hole facing outwards after wax elimination. The investment is then burnt bone white.

When the investment has burnt bone white and the mould cavity shows dull red when viewed through the sprue hole, the burnout is nearing completion. After a suitable period of time, depending on the size of the flask, the furnace controls are adjusted and the mould temperature allowed to fall to about 650°C. At or around this temperature casting may now proceed.

Casting by air pressure
The flask is placed on a heat proof surface and the previously calculated amount of clean, dry casting

Fig. 32. Melting the casting grain
in the crucible depression.

grain, or scrap silver, put in the crucible depression
(Fig. 32).

Fairly rapid heating with a reducing flame is
recommended until the metal fuses together in the
form of a single globule, and takes on the appearance of mercury. Should the metal appear dull or
drossy in spite of the reducing flame, a pinch of boric
acid powder will clean it and improve the fluidity
of the melt.

When the metal appears to spin in the crucible depression, the previously primed pressure caster is
quickly positioned on top of the flask and pressed
firmly down (Fig. 33). The air pressure released will
force the molten metal into the finest detail of the
mould cavity in a fraction of a second.

Hand pressure on the caster should be maintained
for at least five seconds, compressing the metal until
it has solidified in the crucible depression.

Centrifuge casting

Prior to fitting the preheated crucible into its shoe,
the centrifuge spring is cocked (Fig. 34). It is wiser to
do it at this stage of the proceedings rather than after
balancing, as constant tension of the springs over
protracted periods will almost certainly weaken
them.

25

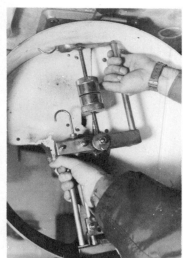

Fig. 33. Using the portable air pressure caster.

Fig. 34. 'Cocking' the centrifuge springs.

When the preheated crucible and its charge of metal is seated securely in the shoe, the hot burnt out flask is clamped into position (Fig. 35).

As in pressure casting, the charge is heated until the silver is fully melted and spinning in the crucible (Fig. 36). A propane torch is illustrated here.

A sharp blow on the centrifuge arm release knob will set the arm spinning (Fig. 37), forcing the molten metal through the hole in the end of the crucible, down through the conical depression in the flask and on into the mould cavity.

No attempt should be made to stop or slow the arm from spinning until it has run its course. Apart from being potentially dangerous, the continuation of the spin consolidates the metal in the casting as it is fed by the residual metal in the conical depression.

26

Fig. 35. Fitting the hot flask in the centrifuge.

Fig. 36. Melting the metal in the ceramic crucible. Note the position of the lower hand, ready to hit the release knob.

Fig. 37. The centrifuge spinning after the release knob has been struck.

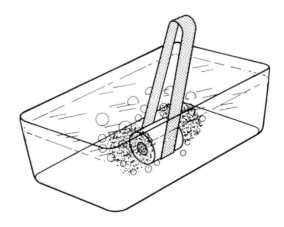

Fig. 38. The investment disintegrating during quenching.

After a cooling period of about five minutes, the hot flask is plunged into a container of cold water (Fig. 38). It is essential that the flask be held as illustrated, for the violent reaction when the plaster disintegrates could be dangerous if allowed to 'erupt' upwards.

Cleaning the casting
Most of the remaining investment may be picked out, using a probe as shown in Fig. 39.

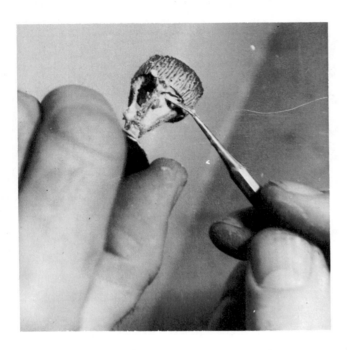

Fig. 39. Picking out remaining investment with a dental probe.

Immersion in a hot sulphuric acid pickle will remove any oxides remaining on the casting. This should be followed by soaking in a solution of sodium bicarbonate and water to neutralise any acid which may have seeped into the metal through the minute porosity holes present in most castings, though not always visible to the naked eye.

A final scrub (Fig. 40), using a mild abrasive powder —a household scouring powder will do—and the casting is ready for the finishing stages.

Fig. 41 shows the finished casting. Any small nodules, caused by air bubbles trapped in the investment slurry, may be picked off using a scraper, or twisted away with a pair of taper-nosed pliers.

28

Fig. 40. Removing investment from the fine detail with a nylon brush.

Fig. 41. The casting after cleaning and pickling.

Finishing the ring

The ring is separated from the sprues by careful use of a piercing saw (Fig. 42). If a vice must be used, the jaws should not be tightened excessively for distortion of the ring will almost certainly occur.

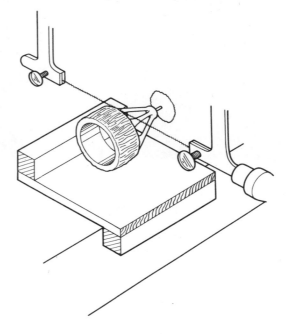

Fig. 42. Cutting away the sprues with a piercing saw.

A pair of end-cutting nippers is often used in the casting trade for removing the sprues and button, but this is just a matter of expediency.

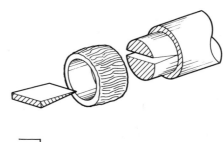

Fig. 43. Holding the ring for facing the edges.

Fig. 43 shows one method of holding the ring whilst the edges are trimmed parallel. A piece of wood dowel is turned down to fit the inside diameter of the ring, whilst a wooden wedge inserted in the saw cut will prevent the ring from slipping when cutting is taking place.

In order to maintain concentricity, the dowel should not be removed from the chuck until both edges of the ring have been faced. Alternatively, the edges of the ring may be filed true.

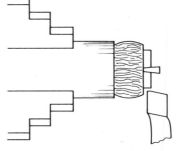

The inside diameter of the ring will need to be brought to the final size. It is now a little less than the original diameter, due to the metal contracting as it solidified from the molten state during the casting process.

At the same time, the inside surface will require the roughness smoothing down prior to polishing.

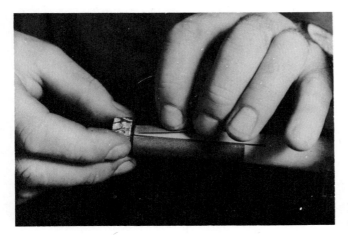

Fig. 44. Fitting abrasive paper on a split wooden mandrel.

Fig. 44 demonstrates a useful way of fastening a piece of fine abrasive paper to a tapered polishing stick. Whilst the dowel is revolving at a moderately slow speed, say 350 r.p.m., the ring is passed to and fro along the abrasive paper (Fig. 45). The ring should be partially rotated from time to time to even out stock removed. Occasional checking of the inside diameter with a ring gauge stick will ensure that the correct amount of metal is removed.

Fig. 45. Smoothing the inside of the ring with the lathe in motion.

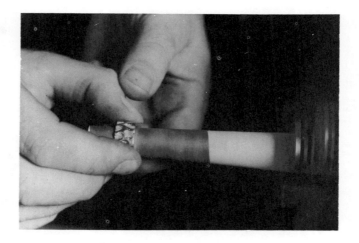

Fig. 46 shows an alternative method of sizing the ring, using abrasive paper on a dowel rod gripped in a vice.

It may be argued that these methods of bringing the ring to size are wasteful of silver and whilst this may be true, the most common alternative method available to the amateur is malleting the ring on a tapered steel triblet until it has been stretched to size. This, of course, would ruin the surface texture.

Fig. 46. An alternative method of cleaning the inside surface of the ring.

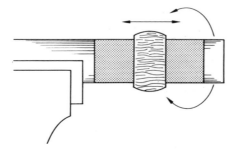

Fig. 47. Brightening a similar ring using a fine crimped steel wire brush fitted to a polishing lathe.

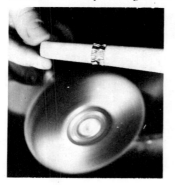

A light scuffing with a crimped steel wire wheel will brighten the surface of the silver (Fig. 47). The use of a tapered stick to hold the ring will prevent it being snatched from the fingers whilst polishing the exterior surfaces. *The use of protective spectacles is advised for this operation, as this type of wheel often sheds a few of its 'bristles' under load.* Fig. 48 shows the same operation using a hand piece on a flexible shaft.

Light pressure on a polishing mop, charged with jeweller's rouge polishing compound, will give the surface a warm lustre (Fig. 49). Undue pressure

31

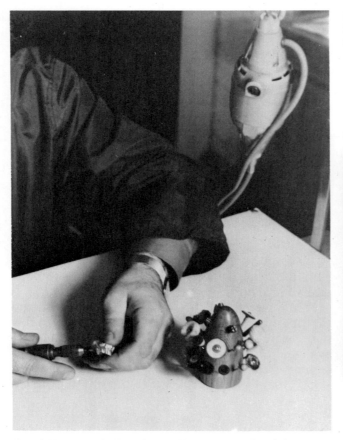

Fig. 48. Using a hand piece and pendant motor for the same process.

Fig. 49. A final polish with a soft polishing mop.

should be avoided as this will cause much of the crisp texture to be smoothed out.

The inside of the ring can be polished with a tapered felt bob charged with polishing compound (Fig. 50). Any polishing compound remaining in the crevices may be removed with the aid of a bristle brush wheel after the rouge has been softened with a suitable metal polish. Alternatively, a light brushing with a jeweller's hand brush charged with a paste of soap and water, will achieve a final lustre. Plate I shows the finished ring.

Note: Should the craftsman wish the ring to be hall-marked, it should be stamped and sent for assay before the polishing stages. The Assay Office will scrape small amounts from various parts of the ring to check the quality of the silver. Subsequent polishing to remove the scraper marks could make it oversize.

Fig. 50. Polishing the inside with a finger bob.

32

3   Design considerations

It is likely and quite understandable, that the new-comer to casting precious metals into jewellery forms will be preoccupied with the technical aspects of the craft. The production of the finished piece complete, without distortions, inclusions, porosity and the hundred and one other calamities that can befall the beginner, will offer sufficient challenge and satisfaction in the early stages. As competence in the technical and craft aspects of casting improves, so will attention shift more and more to questions of design.

For convenience one may classify the factors affecting the designer as technical, functional, and aesthetic. In one sense, the technical constraints are the most important, just because they are the most compelling. Once satisfied, however, they fade to insignificance beside the functional aspects of the problem and in turn once these have been met, aesthetic considerations predominate. It is no accident therefore that the beginner will be principally concerned with mastering the technical aspects of the craft, while the experienced worker will concentrate upon the artistic.

These three aspects are not totally unrelated. For example, one may for aesthetic reasons wish to produce a particular texture and the problem and possibilities are thereafter technical. Similarly, something as functional as a setting for a stone will introduce an aesthetic element into the finished piece. There is in fact a constant interaction between these three aspects of designing and creating jewellery.

Over and above these considerations will be the background of the individual craftsman. Those who approach jewellery from the relative inspirational freedom of the graphic arts, must inevitably undergo a considerable disciplining at the hands of the technical and functional constraints of the craft. Their inspired and ingenious designs may be beyond their powers of execution and unwearable into the bargain.

The newcomer to the craft whose background traditions are those of precision work and mechanical processes, will experience problems of a complementary character. His work is likely to be marked by a lifeless, geometric and over-symmetrical character. For him the problem is to break away from the straight line, the mechanical curve and the simple proportion into the organic shapes and proportions that are living and delighting to the eye.

Technical aspects of design
Almost any shape or size of pattern that can be formed in wax can, in principle, be cast by the lost wax method. However, when using the centrifugal and pressure casting techniques available to the amateur or small scale producer, there is in practice, a fairly severe limit to the size and shape of pattern that can be accommodated in the normal casting flask. Similarly, the size of the available crucibles may set a limit to the volume of metal that can be employed in any one job. When large items are being produced, it may be necessary to break the pattern down into segments that can be cast separately and subsequently reassembled, by soldering, riveting or some other technique.

The first tentative ideas may be set down in a sketch book and worked up in graphic form before any attempt is made to create the pattern. Some craftsmen, however, prefer to design and experiment directly in the medium in which the pattern is to be made, be it wax or metal. Sketches take relatively little time and effort so that for the beginner at least, it seems worthwhile getting a few ideas down on paper to begin with, even if they are subsequently modified in the course of constructing the pattern.

The decision in regard to the medium in which the pattern is to be made depends on a number of different factors. If equipment for reproducing wax patterns by the injection of molten wax into a vulcanised rubber mould, is available, then the original model may be made in any suitable metal such as silver, tin, copper, or aluminium.

Where this equipment is not available or when the

18ct gold leaves with diamonds by Andrew Grima.

piece is intended to be a 'one off' job, then the original can be created directly in wax. If necessary, the casting produced from this wax model can be used as the master for making further wax models, using the wax injector and a vulcanised rubber mould.

Thus, where facilities for reproducing wax patterns are available, the designer can please himself as to whether he works in wax or metal. When these are not to hand, he must work directly in the medium of wax.

For most purposes, the best original patterns are made in metal. Smooth polished surfaces and hard firm lines can be produced in metal, in a way that is not possible when working in wax. The tools and techniques are those familiar to the metal worker and far greater precision can be achieved where this is desired. Also with the metal being far more resistant to the tool, much finer detail is possible and its greater strength permits thinner sections to be produced. This is particularly important when making claw settings for stones and delicate ornamental designs.

There are some circumstances, however, where it is best to create the original pattern in wax. Its soft malleable properties impart a gentle roundness to the design that harmonises well with the richness of the precious metals. Heavily undercut designs may have to be worked in wax owing to the difficulty of extracting the wax pattern intact from the vulcanised rubber mould. Also by working in wax the designer can take advantage of various techniques not possible with metal. Pointed metal scrapers can be used to produce a variety of surface textures, designs, and reliefs. Gently heated in a clean flame, doming punches, chasing tools and other metal implements can be used to fashion the model. A wide variety of knives, scalpels, and scribers can be used to carve the wax to the exact form required.

Molten wax can be applied like paint with a brush or spatula and brooch and ring settings can be built up about the stones they are to hold. Surplus wax is carved away and the stone is removed from the wax claws once they have been made flexible by immersion in warm water. Molten wax poured gently into cold water produces fantastic convoluted shapes that can be translated in every minute detail into gold or silver. Water soluble wax used in conjunction with the normal kind, can produce hollow forms not possible by other means.

35

Particularly useful for working wax are an electric wax pen and spatula. The pen extrudes a continuous or intermittent worm of wax and can be used in much the same way as an icing syringe, to build up designs. The electric spatula is used for carving and joining.

Though incredibly fine detail can be reproduced by casting, there are many occasions when the piece will be partly cast and partly constructed. For example thin box sections are best constructed from sheet metal. Fine claw settings can be added to the cast body of the ring, brooch or earings. In fact, findings of all sorts such as joints, catches, pins and hinges are for the most part better added by soldering them to the main body of the piece rather than the attempt made to cast them in situ.

Textured gold brooch with sapphires and diamonds by David Thomas. From the Goldsmiths' collection.

Functional aspects of design

Jewellery functions at a variety of levels. It may be magical, mystical, symbolic, representational, practical, ornamental, ostentatious, or just a financial investment. Whatever else it is, it must be wearable.

Few people are totally insensitive to the magical and mystical aspects of jewellery, and the designer who studies the history, religions and mythologies of his own and other cultures, will find in them a rich and profitable source of artistic stimulation. Symbolism enters the design of much jewellery in forms as varied as military and heraldic insignia, crosses, peace symbols, hearts, eternity rings, masonic marks, and club badges. More occasionally, the designer is set the task of direct representation in his work, as

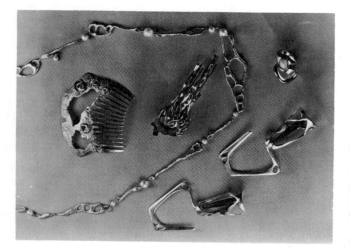

Silver comb, silver and pearl necklace, silver wedding and engagement ring with turquoise, silver gilt and tourmaline brooch, silver and carnelian ear-rings with removable cages by Dorothy Low. From the Goldsmiths' collection.

Quartz crystal brooch in a circular
gold setting by John Donald.
From the Goldsmiths' collection.

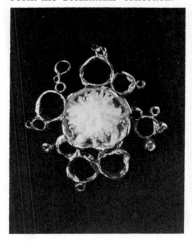

for example, in making bracelet charms, cameos, and
facsimiles of natural and man-made objects. In this
context the investing and burning out of natural
objects such as flowers, seed pods and leaves, is a
particularly useful technique. Usually, the design is
more effective though, if it is an abstracted form of
the original that it is to represent.

At a practical level the different items of jewellery
must do the job for which they are intended. Hair
combs and ornaments must be secure and control
the hair. Earrings, brooches, clips, tie tacks and cuff
links must be secure and not injure the person or the
clothing of the wearer. Similarly, buckles, belts,
chains, lockets and most other items must have certain very practical characteristics incorporated in the
design if they are to be considered at all functional.

So far as ornament and ostentation are concerned
these aspects of the function of jewellery are largely
matters of aesthetics and will be considered below.
The skilful craftsman will, however, use his technique to create the maximum effect with the available materials. In casting this is very much a matter
of using hollow, thin sections where possible.

Jewellery as an investment, or a hedge against
inflation, really demands expert advice especially in
regard to the value of stones. Just occasionally, if he
is fortunate the designer will work for a client who
wishes to spend a substantial sum of money. The
pity is that if the piece is of sufficient value it will,
for reasons of security, be unwearable and will spend
the greatest part of its existence in a bank vault!

Jewellery must also function in the sense of being

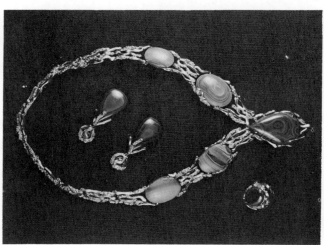

Necklace, matching ear-rings and
ring in 18ct gold with
malachite by John Donald. From
the Goldsmiths' collection.

37

wearable. It must fit comfortably and securely without unduly restricting the movement of the wearer.
It should not be too heavy, or fragile, and sharp
spikes or angles should be avoided, where they might
cause personal injury or damage to clothes. All this
is no more than common sense, but the beginner
will almost certainly find that some of his more

A selection of ear-rings in 18ct
gold, one pair set with drop
shaped tourmalines and the other
with cabochon emeralds,
by John Donald.

extravagant designs are less popular than he hoped
for reasons as obvious as these.

Recent developments in jewellery fashion have
eased these strictures somewhat. Rings are now
commonplace that would have been regarded as unwearable only a few years ago. In the current
fashion for pendants, objects the size of saucers are
worn by both men and women. Jewellery is more
and more being designed for wear by men as well
as women and not just in the traditionally acceptable
form of rings, tie-pins and cuff-links, but also in the
form of pendants, necklets, bracelets and even ear-
rings.

38

A selection of 18ct yellow gold
watch bracelets by John Donald.

This trend seems likely to continue and to provide
opportunity for the designer to show his originality
and flare. A further reflection of the more liberal
attitudes to jewellery fashion, is the appearance of
body sculpture, ornamental belts, rings and bracelets
combined, anklets and the wearing of rings by the
half-dozen.

The modern designer is fortunate to be working at
a time when anything goes; when any original piece
stands a good chance of finding someone to wear it.

Aesthetic aspects of design

All forms of communication proceed by arousing in
the receiver, perceptual and conceptual responses
that have been developed within him in the course
of his previous experience. A person's ability to per-
ceive beauty is acquired and in the process of acquisi-
tion, the individual also learns what his particular
culture is prepared to accept as beauty. One cannot
therefore define what is eternally and universally
beautiful, for each age and culture sets its own
standards.

Psychologically, perception is the search for patterns that can be related to, and thus assimilated by,
existing mental structures. This assimilation of the
new in terms of the old, or the unfamiliar in terms
of the familiar, is a satisfying experience. One finds
an object beautiful when it is sufficiently similar to
previously encountered objects of beauty, to permit
successful recognition and assimilation by existing
perceptual and conceptual structures. At the same

39

time it must be sufficiently novel to sustain interest and attention by demanding from the viewer some effort after form and meaning.

The process is a two-sided one. The new stimulus is assimilated by the established structures and produces in those structures, subtle changes. These changes represent the cumulative effect of experience or learning. In this way, by extending a person's visual experience of beauty, you increase his capacity for perceiving it in new situations.

All that has been said of perceptual processes is related to the processes of creative designing. Creativity is often spoken of as though it were a faculty, either possessed or not, that enables the individual to generate new and unique ideas. Without wishing to deny innate individual differences, it is a fact that, productivity in design depends to a great extent on prior consumption in terms of perception. The creative artist cannot work in an experiential vacuum. He may be unique in the speed with which he masters and absorbs the ideas of his predecessors and contemporaries, and in the range and originality of his innovations. But the basis upon which any creator must work is the wealth of his own visual experience.

It follows then, that the original designer, while he will not blatantly copy the work of others, will be a student of it. Many of the great innovations represent the fusion and harnessing of ideas from sources right outside the field in which the designer himself is primarily interested. He will, therefore, seek ideas and inspiration at any time, in any place and in any field of knowledge. The methodical will record these ideas in sketch books; the more disorganised will rely upon unaided memory, but any designer will saturate himself in visual experience and will work to extend his perceptual structures.

There is a popular contrast drawn between convergent and divergent thinking in discussion of creativity. Convergent thinking is said to be involved in the production of a single and possibly unique solution to a problem. Divergent thinking is said to operate when the problem allows of many possible solutions all equally valid. In the former the emphasis is upon critical evaluation of any idea put forward in the light of the stringent demands of the problem. In the latter, the stress is upon the range and variety of the possible solutions offered and the

A heavily textured gold ring with Gotham synthetic crystals.
Courtesy of Technical specialities International Inc.

fluency with which they are produced.

The convergent/divergent distinction does have a certain relevance to the situation of the designer of jewellery. Set the problem of producing a design, for say a brooch, ring, necklace or whatever, there are limitless possible solutions to the problem. The need then, is first for divergent thinking—for fluency, variety and originality of ideas. At this stage there should be an explosion of the mind. Ideas are needed as fast as they can be drawn. Nothing should be rejected (Fig. 51).

Later, when the flow of ideas subsides, evaluation and modification take over, as the more promising ideas are subjected to scrutiny in the light of constraints imposed by technological, functional and aesthetic considerations. The modification and evaluation process continuously applied, may bring

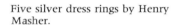

Five silver dress rings by Henry Masher.

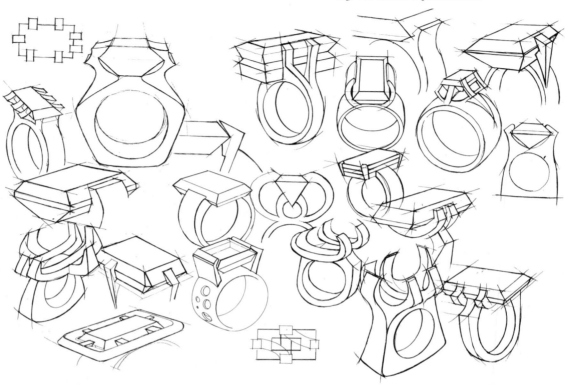

Fig. 51. Creative possibilities.

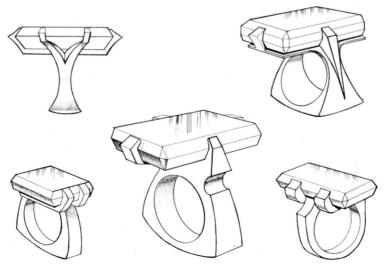

Fig. 52. Narrowing the field.

Fig. 53. Sketching the final choice.

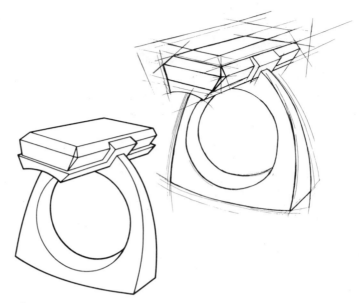

Fig. 54. The finished design sketch.

Silver bracelet with turquoise cabochon stones by Caroline Couchman. From the Goldsmiths' collection.

about an acceptable solution or recourse may be necessary to unfettered production once more.

As one or two promising ideas begin to emerge these should be worked up into more polished drawings, Figs 52–54. Where necessary, other views should be drawn along with enlargements of critical details such as catches, clasps, hinges, and settings for stones.

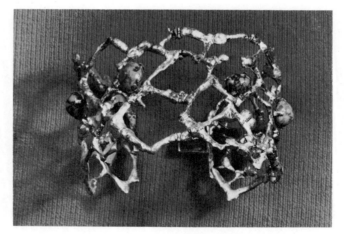

Individuals differ in their ability, in regard to the production and evaluation aspects of the process. Some find the greatest difficulty in producing an easy flow of ideas. They seem so critical of their own efforts and fearful of the scornful comments of others, that the ideas are rejected before they are

Silver cuff-links by Keith Redfern. From the Goldsmiths' collection.

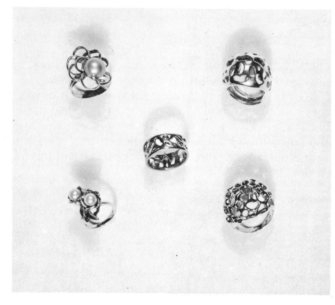

Five silver rings by Sunny Erwin. Courtesy of Kerr Manufacturing Co.

Sea horse brooch in sterling silver by Leslie Durbin. From the Goldsmiths' collection.

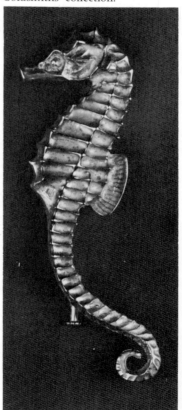

coherently formulated. Compounding the problem is usually a limited ability to draw. This, again, is often the product of a lack of practice induced by an over self-critical attitude. Essential to this part of the process is self-confidence and the ability to tolerate failure. If this can only be achieved in the absence of an audience then design in private. Success will come in time and there will be a cumulative growth in competence and self-confidence. Meanwhile, seize every chance to draw, sketch, or just doodle. Ideas will soon begin to emerge on paper and even if they are not immediately relevant, they won't be wasted.

44

Six gold rings with semi-precious stones by David Pearce. From the Goldsmiths' collection.

Cuff-links by David Pearce. From the Goldsmiths' collection.

Cuff-links in gold by John Donald. From the Goldsmiths' collection.

Quite as damaging to the creative process as lack of inspiration, is lack of rigour in evaluating the ideas produced. If the design cannot be executed in wax, then the piece cannot be made by the lost wax method. If when it is finished it proves too cumbersome, insecure or just plain ugly, then it is no more successful than if the idea had been still born through an unduly self-critical attitude.

Where rigorous thinking is lacking, then call in a critical friend whose tact, technical competence and taste you can rely upon. Alternatively, draw up a checklist of criteria and ruthlessly subject each idea to it.

Creativity is sometimes confused with the divergent or productive thinking aspect of the designing process. Fluency is not enough, the creative craftsman needs ideas in plenty, but he needs to be able to pick out the good from the bad. Moreover, he needs the technical skills and application to follow the ideas through to a successful completion. If he hopes to earn his living by his creative effort, he will also need that other set of skills that will allow him to survive in the market place, for it is there that many creative jewellers come to grief.

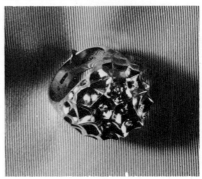

Gold ring with diamonds set in craters by David Thomas. From the Goldsmiths' collection.

Six gold wedding bands by Jocelyn Kingsley. From the Goldsmiths' collection.

4   Pattern construction

Materials

Almost any material which can be consumed or 'lost' by normal burnout temperatures may be used for pattern construction. Whilst wax remains the most common pattern material, 'found' or natural objects such as leaves, seed pods, tree bark, insects and soft crustacea offer interesting possibilities.

Few, if any, manufacturers are prepared to offer materials on the open market until they are assured of a regular demand. This is particularly true in the case of non-commercial waxes for jewellery casting and the relatively expensive American materials appear to be the only ones available which are intended for this particular application.

This then is the current situation but it may well improve during the next decade as investment casting becomes more popular.

Fortunately the amateur does not have to rely solely on expensive imported products, as a wide variety of waxes intended for dental use is currently available at modest cost and will, in the majority of cases, offer excellent alternatives.

The tables on pp. 48 and 49 give an indication of some of the waxes available which are suitable for hand forming techniques of pattern construction.

Type	Characteristics	Form of Supply	Suggested Applications
Sheet Casting Wax	Thin, smooth, and transparent. When warmed it may be formed into a variety of shapes.	Boxed sheets, $3'' \times 5\frac{1}{2}''$, in various gauges. Two types are available, green and pink, the latter being slightly more rigid.	May be melted and cast into slabs, coated on ring mandrels, cut into strips for ring bands, or pierced with a heated modelling tool.
Jeweller's Wax Wires	Flexible but strong. May be twisted or bent to form a desired shape.	Boxed 4″ lengths in assorted gauges. Round, half round, square and rectangular sections.	Particularly useful for 'open-work' pieces of filigree nature, or claw setting for gem stones.
Jeweller's Sprue Wax	A flexible wax which is formulated to melt out at a lower temperature than other waxes.	Boxed 4″ lengths in various gauges.	Mounting a pattern on the crucible former prior to investing.
Jeweller's Carving Wax	A green hard wax which carves easily with minimal flaking and chipping.	1-lb blocks, $3'' \times 7'' \times 1\frac{1}{2}''$.	Useful for thicker 3-dimensional carved forms. It may also be melted and cast into thinner slabs prior to shaping, or used for wax build-up.
Jeweller's Master Pattern Carving Wax	Extremely hard. It may be sawn, carved, filed, drilled, or even turned on a lathe. Red in colour.	1-lb blocks, $7'' \times 3'' \times 1\frac{1}{2}''$. Solid or sliced in an assortment of pieces from $\frac{1}{4}''$–$2''$.	Easier to work than metal, it is useful for making a preliminary pattern which, when cast and finished, will be used in conjunction with a vulcanised rubber mould for 'repeats'. Can also be used for one-of-a-kind patterns.
Jeweller's Master Pattern Ring Tubes	As for previous wax.	Available in various round centre-hole tube, round off-centre tube, D-section tube and solid bar.	As for previous wax.
Wax Pen Refill Waxes 1. Avocado Green	Regular jewellery grade, tough and slightly brittle.	$\frac{1}{4}$-lb boxed sticks to fit wax pen.	General purpose wax.
2. Fabufill Red	A less brittle, lower melting point wax.	As above.	Its lower melting point allows rapid and easy build-ups.
3. Pink Ultraflex	Highly flexible and tough when cool.	As above.	Useful for delicate wax build-ups, allows patterns to be drawn in the air.

DENTAL WAXES

Type	Characteristics	Form of Supply	Suggested Applications
Modelling Wax (Toughened)	Transparent and flexible, coloured red to pink.	1-lb boxed sheets 7" × 3⅜" in various gauges.	A general purpose wax, it may be warmed and bent into three dimensional forms, melted for mandrel coating or cast into slabs for carving. Its transparent nature allows two dimension forms to be traced direct from a drawing.
Model Cement (Sticky Wax)	Hard and very brittle under normal conditions, when heated it is viscous and extremely adhesive. Creamy white in colour.	½-lb boxed sticks.	Useful for tacking two or more pieces of wax together prior to a more secure joint being formed.
Inlay Wax	Hard and brittle under normal conditions, when warmed it becomes easily mouldable. Usually blue or violet in colour.	Boxed strips or blocks.	May be used for build-up prior to carving.
Profile Wax	A flexible green wax with a low temperature melting point.	Boxed in 5" lengths in a variety of sections and sizes. The round section is also available on 250 gm reels in diameters from 2.5 mm to 5 mm.	Ring bands, sprue wires, claw settings, etc.

COMMERCIAL JEWELLERY WAXES

Type	Characteristics	Form of Supply	Suggested Applications
Injection Wax	A tough wax, viscous when fully melted, its generous plastic range allows patterns to be easily removed after injection into rubber moulds. Available in a variety of colours.	Sold by weight in slab or pellet form.	Injection into rubber moulds for repetition work. May be added to melted modelling wax to impart hardness. Particularly useful for 'accidentals'.
Water Soluble Wax	A tough wax which dissolves in water. Should be stored in air-tight bags.	Sold in slab form by weight.	Useful for forming hollow cores.
Wax Ring Patterns	Fairly thick sections, made from tough injection wax.	A fairly wide range of styles and sizes.	May be carved or built up to suit individual taste.

Tools and equipment

Pattern construction by hand techniques requires only basic tools in order to achieve a variety of shapes and textures. Initial outlay need only be minimal and many of the tools may be fashioned from odds and ends or adapted from existing items.

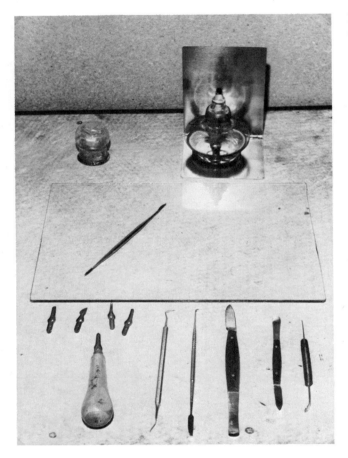

Fig. 55. Basic equipment for wax carving and forming.

A flat wipe-clean surface is essential and a piece of sheet aluminium about 300 mm square, or a piece of stout glass, having had the sharp edges rubbed smooth with a coarse carborundum slip stone, are both ideal for the purpose.

Fig. 55 illustrates a small alcohol or spirit lamp standing on a piece of aluminium sheet which has been bent at right angles to form a useful shield against draughts. A coat of matt blackboard paint applied to the inside of the angle will minimise reflected glare from workroom lights should it prove troublesome.

Other items shown in Fig. 55 are a lino cutting tool with an assortment of blades, stainless steel dental wax carvers and wax knives. A close-up of some modelling tools is shown in Fig. 56 where the tools on the left of the picture have blades utilising an assortment of darning needles and a piece of 1.6 mm welding rod bent at right-angles. All the blades have been inserted into wood dowel handles, allowing good control and heat insulation.

It is important that all modelling tools are kept clean and bright during use. Periodically they should be passed through the flame and the surplus wax wiped away with a piece of cloth specially reserved for this purpose.

A bunsen flame may be used instead of a spirit lamp though it does have certain disadvantages.

Fig. 56. Wax modelling tools.

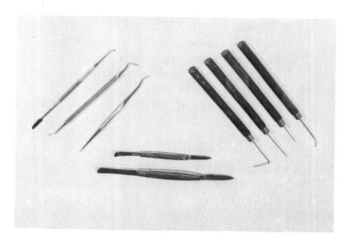

Unless it is specifically designed for wax work, because of the higher flame temperature there is a strong possibility of the modelling tool being overheated. This gives rise to carbon deposits on the blade and loss of control, where radiated heat causes thin wax areas to melt too rapidly. A gas flame will also mean that the work area must be sited near to available gas points, which may not always be convenient.

Modified miniature soldering irons with an operating temperature of 100°C are available. This temperature is a little too high for some applications, though

51

the inclusion of a suitable rheostat or a solid state controller (such as the type used for varying the speed of portable electric drills) in the circuit will allow more control over tip temperature.

Fig. 57. A precision electric wax knife with a variable temperature control. Courtesy of F. & H. Baxter Ltd.

A purpose-built electronic wax modelling tool with a variable temperature control is shown in Fig. 57. It is supplied with a variety of different shaped tips, though others to suit particular requirements may be easily made from copper or brass rod. The range of tips, coupled with the variable temperature control, make this tool extremely versatile, though a certain amount of practice is necessary before its full potential may be realised.

The wax pen and spatula illustrated in Fig. 58 is an expensive and sophisticated American importation. Originally conceived in 1966 as a professional dental waxing unit, it has a control unit with individual temperature control to both pen and spatula.

The pen, containing a charge of solid wax rod, is electrically heated and the flow of molten wax through the tip orifice is controlled by the lever mechanism clearly shown in the illustration. The spatula with its various shaped tips, is used for carving, shaping and smoothing the wax model. Although extremely versatile in its potential application, at a price of around £65 and with wax refills in the region of £3 per ¼ lb, the cost/effectiveness of this unit would need to be seriously considered by the amateur.

Fig. 58. Electric wax pen and spatula. Courtesy of Hoben Davis Ltd.

Hand forming techniques

Use of mandrels

Aluminium mandrels conforming to standard ring sizes are virtually indispensable when making wax ring patterns. Not only will they support the work during carving or build-up operations, but also their use obviates the tedious and sometimes tricky task of drilling or carving a hole of the correct finger size in a wax slab. This type of mandrel is preferable to the triblet type which, by nature of its shape, will form a taper on the ring shank.

Wax may be built up in several different ways. A strip of sheet modelling wax, or 'D' section profile of suitable length may be bent round the mandrel, allowing a small overlap, and a warmed wax knife used to trim away the surplus. The joint is fused or welded with a heated modelling tool.

As described in Chapter 2, a coating of wax may be built up by rotating the end of the mandrel in a container of melted wax until the desired thickness is obtained. When set, the wax may be turned or carved. These two methods are particularly suitable for the construction of wedding bands.

Where shapes other than circular are required, the mandrel is placed on a sheet of aluminium within a surrounding plasticine wall of suitable form and height. The reservoir formed is then filled with liquid wax.

Surface shrinkage cavities may occur where the wax depth exceeds 18 mm. However, a heated spatula or modelling tool will soon remelt and level off the irregularities.

Flow-lines or creases caused by the rapid setting of the liquid wax when it first comes in contact with the cold mandrel may be minimised if the latter is warmed before the film of oil is applied.

Melt-outs

This technique is particularly useful for making pendants and brooches from sheet modelling wax. A heated needle or spatula is pushed through the sheet and the melted wax quickly blown out. The tool is reheated and the edges of the melt-outs are formed and smoothed as desired (Fig. 59). The transparent nature of the wax will allow a form to be traced on the wax direct from a drawing (Fig. 60).

53

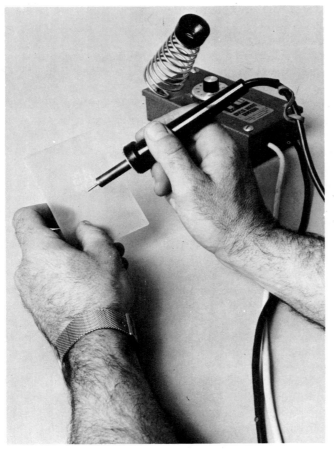

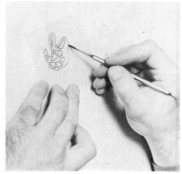

Fig. 59. Melting out with a modified soldering iron.

Fig. 60. Tracing a form on transparent modelling wax.

Fig. 61. Wax 'accidentals'.

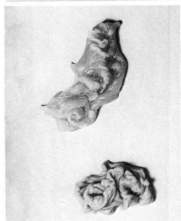

Hollow cages may be constructed by first warming sheet wax under a lamp or in warm water and moulding it over a suitably shaped pebble, prior to melting out. It will, of course, be necessary to make the piece in two parts which may be subsequently joined by fusing or welding with a hot probe.

Free-form or accidentals

By pouring melted wax into ice cold water, unique convoluted shapes may be created (Fig. 61). There is no set technique and one must experiment with varying depths of water and different pouring heights. The type and temperature of the melted wax too, can have an effect on the resulting free-form.

Once the outer layer of the poured wax has set,

the piece should be lifted out and the surplus wax poured away. This will economise on the amount of metal required for the casting. Alternatively, water may be splashed on to the back of the piece whilst the latter is still molten, forming interesting textural cavities.

The shape may be cast 'as formed' or carved and textured with modelling tools as desired. Short lengths of profile wax may be welded in position to form claw settings for gemstones, or bails for necklace chains, thus avoiding subsequent soldering operations.

Two unusual pieces of 'free-form' jewellery, illustrating the potential of this technique, are featured in Figs 62 and 63.

Wax build-up

More wax may be added to a pattern by heating a modelling tool—the right-angled tool in Fig. 56 is ideal for this purpose—touching the tool into scrap wax to melt and gather some, and depositing on the pattern where the build-up is required. Larger amounts are best deposited with a heated spoon-type spatula.

Almost any type of suitable wax may be used for build-up purposes though if delicate detail is to be subsequently carved, then one of the harder waxes, such as injection or carving, should be used.

Granules

Small wax balls, or shot, up to about 3.2 mm may be easily formed by gathering wax—the injection type is most suitable—on to a heated modelling tool or spatula and then touching it down on to a piece of clean aluminium sheet. The shot may be tacked in position on the pattern with model cement (sticky wax) on a heated needle and securely fixed by fusing or welding.

This method was used to construct the 'setting' for the ring illustrated in Fig. 64.

Texturing

A variety of textures may be applied to wax by means of heated needles, knife edges, spatulas or pieces of odd scrap metal filed to a given shape. Where quite deep impressions are to be made, an electrically-heated tool having a variable temperature control will be found to be the most suitable for

Fig. 62. Sterling silver pendant by Alan Alderwick, former student of Worcester College of Education.

55

moulding the wax without melting it.

Rough textures may be formed by scuffing the surface of the wax with a small steel wire cup brush fixed on a small diameter mandrel and used with a suspension unit. Alternatively, a fine spray of liquid wax may be applied to the pattern with a mouth-blown artist's atomizer spray.

The wax pen

The wax pen is an electrically-heated hollow pen with an intricate flow system. A wax rod is inserted in the upper part of the pen and the unit switched on by the control on the consol.

A certain amount of experiment will be needed to ascertain the optimum temperature setting for each of the three recommended waxes. Once discovered, the settings should be noted for future reference.

At the lower end of the pen is a lever which controls the amount of wax allowed to pass through the orifice and a trigger to turn the flow off and on.

Ten to fifteen minutes should be allowed for the wax to melt and any temptation to move the pen controls during this period should be firmly resisted. Damage to the delicate mechanism could easily result if force were used.

Should the pen be inoperative after the recommended warm-up period, then in all probability the air vent at the end of the pen will be found to be clogged with wax. This will have been caused by the pen being held point uppermost so that the molten wax fills the vent hole. A fine heated needle or piece of wire may be used to clear the vent, though in extreme cases the unit may have to be dismantled.

Although only three basic applications are illustrated (Figs 65–7), in conjunction with the spatula and by dint of using different waxes and varying the thickness of flow, a fantastic variety of forms may be readily created. A word of warning though, the unit, because of its relatively high cost and delicate nature, should be handled and stored with more than the usual amount of care.

Pattern finishing

Because of the excellence of surface detail offered by investment casting, the surface finish on the pattern is of considerable importance if one is to minimise subsequent filing and finishing operations.

Fig. 63. Sterling silver brooch by Alan Alderwick, former student of Worcester College of Education.

Fig. 64. A 9ct gold dress ring by the author.

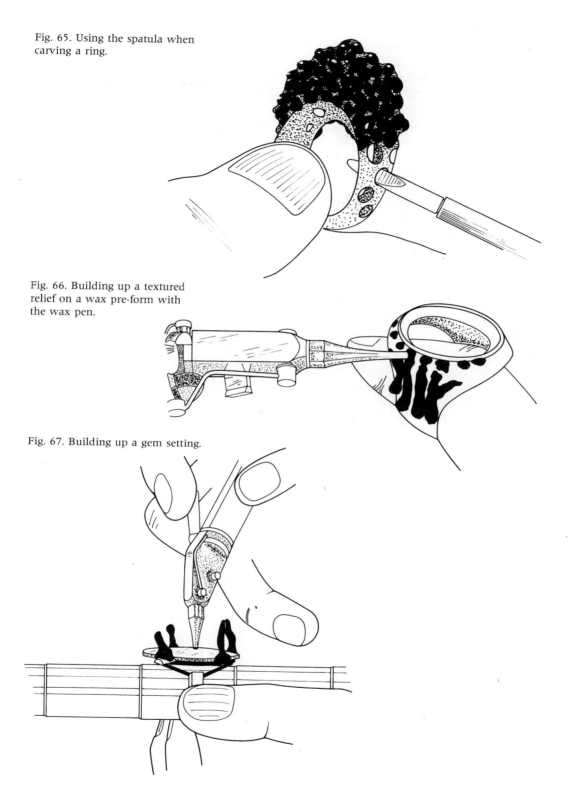

Fig. 65. Using the spatula when carving a ring.

Fig. 66. Building up a textured relief on a wax pre-form with the wax pen.

Fig. 67. Building up a gem setting.

Any irregularities should be carved or smoothed away and the pattern cleaned of all grease and dirt by gently scrubbing with a soft brush charged with a bland soap or liquid detergent. A final polish may be imparted by quickly passing the pattern through the flame of a match or spirit lamp, though this method is not without some risk to the piece. A safer method of polishing is by gently rubbing the pattern with a piece of water-moistened cotton wool.

Commercial methods of pattern production

It would not be economically viable for a trade caster or manufacturing jeweller to make single patterns and cast on a one-off basis, so a method of producing identical patterns quickly and cheaply had to be devised.

The most widely used method is by rubber mould reproduction. The pattern is first made in metal, usually silver, which may subsequently be rhodium plated to avoid sulphur contamination from certain types of mould rubber. Where final size is fairly critical, as in rings, an allowance of between 5% and 10% is made to allow for the accumulative contraction of the rubber, wax, metal, and the cleaning and polishing of the completed casting.

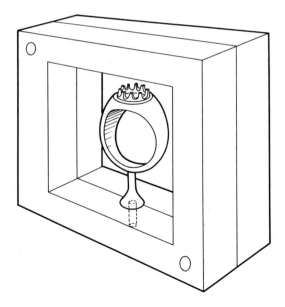

Fig. 68. A ring pattern mounted in an aluminium mould frame.

A piece of small diameter rod, sometimes fitted with a conical end, is soldered to the pattern to form the wax sprue (Fig. 68). Where the pattern is of the

58

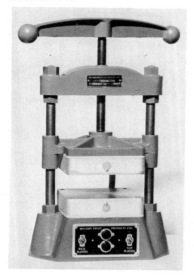

Fig. 69. An electric vulcanising press for the production of rubber moulds. Courtesy of William Frost Products Ltd.

open-work type as illustrated by the nurse's belt buckle in Fig. 79, two or three extra wires may be linked from the pattern to the central sprue. This will ensure a more rapid and even flow, filling the mould in the waxing and casting operations.

The surface finish on the pattern is of considerable importance as even the most minute blemish will be faithfully reproduced on the wax facsimile and subsequently on the finished casting.

The first stage of the process calls for a suitably sized aluminium mould frame which is placed centrally on the lower platen of an electrically heated vulcanising press similar to the type illustrated in Fig. 69 and pieces of uncured sheet rubber laid within the frame until the half-way mark is reached.

The pattern is then laid on the rubber and the central metal sprue located in the hole provided. More rubber is then packed around and over the pattern until the whole frame is filled. The platens are brought together by turning the winding arm at the top of the press.

It is essential that pressure is applied gradually. Full pressure should be applied only after the rubber has had time to warm through. Uncured rubber expands considerably during vulcanisation and if full pressure is applied immediately the further pressure exerted by the rubber could, in extreme circumstances, fracture the press frame.

The vulcanising temperature is in the region of 150°C and once the platens have reached this temperature, the indicator lights, being in circuit with the thermostats, switch on and off automatically when maintaining the pre-set temperature.

Provided the platens have been preheated before the insertion of the mould frame, the time required to fully vulcanise the rubber will be between 45 and 90 minutes depending upon the frame thickness. A good rule-of-thumb method is to allow 15 minutes for each 6 mm of rubber.

Some casters drill a few small holes through the wall of the frame, along the centre line. This is by no means a bad plan for not only will it indicate the stage of vulcanisation reached when the 'worms' of rubber squeeze out, they also permit the excess rubber to escape, thus minimising subsequent warpage of the mould due to over-compression.

A simpler method of vulcanising, particularly suited to amateur requirements, may be achieved by using a simple press (Fig. 70) in conjunction with an ordinary domestic oven, though the same rule of applying pressure gradually should be observed.

There are two other methods of mould making which require the minimum of equipment. One calls for the use of a fairly new material of American origin called See-Thru Compound. This is a milky white liquid which is poured into a special u-shaped mould frame, clamped between two pieces of glass with strong spring clips (see Fig. 71).

Fig. 70. A simple vulcanising press for amateur use. Courtesy of Hoben Davis Ltd.

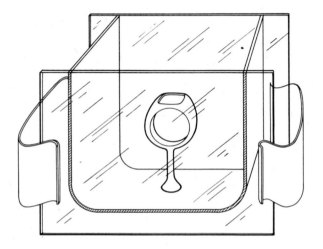

Fig. 71. An aluminium and glass mould frame for 'See-Thru' compound.

The mould should be subjected to vacuum prior to baking at 150°C in a domestic oven. When fully cooked it will be flexible and transparent, which makes subsequent cutting to remove the pattern a much easier proposition for the beginner.

Liquid silicone rubber that cures at room temperature, obtainable in several different grades, is another excellent material for mould making. The use of a catalyst is necessary and, depending upon the type and proportion used, curing time may be varied between 28 minutes and 48 hours.

A much lighter and simpler mould frame may be used with this material. Some workers use frames constructed with LEGO type plastic building bricks, whilst others find a 'dam' of plasticine quite satisfactory for most patterns. The liquid compound is poured alongside the pattern into the mould and is set aside to cure for the recommended period.

When the mould is vulcanised or cured, as the case may be, it must be cut into two parts so that the original metal pattern can be removed. This, depending upon the form of pattern encapsulated, may be quite a tricky operation, calling for the use of razor-sharp narrow-bladed knives and some forethought.

Locks, clearly shown in Fig. 78 are cut into the rubber so that the two parts of the mould may be correctly aligned when the liquid wax is injected into the mould cavity.

Fig. 72. A piece of aluminium foil, fitted at the half-way stage, will make mould cutting much easier. Note the cut-outs which will form the location notches.

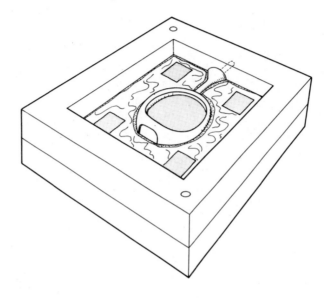

Fig. 73. A third hand helps to separate the mould during cutting.

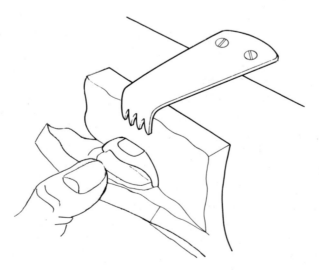

61

Apart from using the 'See-Thru' compound, the beginner may find cutting easier if, during the rubber packing stage prior to vulcanising, a piece of suitably cut aluminium foil is laid over the pattern at the halfway stage (Fig. 72). This will form a parting line almost up to the pattern.

A single or multiple pronged piece of metal, or even a beer or soft drink can opener screwed to the bench, will help to separate the mould during cutting, as shown in Fig. 73.

With the metal pattern successfully removed, the mould cavity may be given a light dusting of french chalk to afford easy removal of the wax pattern, and the two parts of the mould re-aligned ready for the wax injection.

In order to prevent localised compression of the pattern cavity, the mould is usually held, sandwich fashion, between two suitably sized pieces of aluminium sheet.

The wax injectors illustrated in Figs 74 and 75 are basically sealed containers with a thermostatically controlled heating element to keep the melted wax at the optimum temperature. They are normally pressurised with an air line or a car foot pump and when pressure is applied to the injector nozzle a stream of liquid is forced out.

The mould is pressed, sprue hole foremost, against the injector nozzle and the wax is forced into the cavity (see Fig. 77). Some commercial casters use at least three identical moulds, so that whilst the wax is setting in two of the moulds the remaining one may be injected.

Fig. 78 clearly shows the wax pattern as the top of the mould is peeled away. It is generally accepted that girls perform this type of work better than their male counterparts. Their nimble fingers are more adept in handling the delicate wax patterns.

Some examples of commercial wax patterns are illustrated in Figs 79–82.

Suggested materials and equipment

Basic
Assorted waxes
Wax knives and modelling tools
Files of assorted cut and section
Modified lightweight electric soldering iron
Assorted sizes of ring mandrels with stand

Fig. 74. A wax injector with two nozzles. Courtesy of Nesor Ltd.

Fig. 75. A modern Italian wax injector of elegant design. Courtesy of Ferraris Engineering and Development Co. Ltd.

Fig. 76. A small wax injector.
Courtesy of Hoben Davis Ltd.

Fig. 77. Injecting the wax into
a rubber mould.

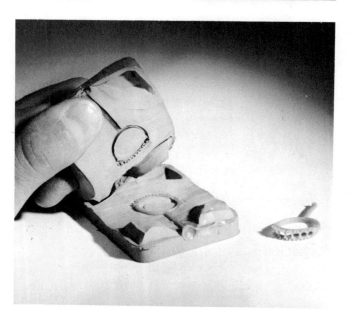

Fig. 78. Peeling the top part of the
mould away to reveal the pattern.

63

Ring gauges and ring stick
Glass or aluminium work surface
Spirit lamp
Tripod stand with wire gauze
Thermometer, 0°–100°C
Container for melted wax
Plasticine
Thin oil
Jeweller's soft polishing brush
Liquid soap or detergent
Cotton wool

Adjunct
Electric wax modelling tool
Electric wax pen and spatula
Metal patterns
Aluminium mould frames
Electric vulcanising press
Wax injector
'See-Thru' mould compound
Aluminium U-shaped frames
Glass plates and spring clips
Vacuum machine
Gas or electric domestic oven
Cold-cure silicone rubber

Figs 79–81. Typical commercial wax patterns.

Fig. 79.

Fig. 80.

Fig. 81.

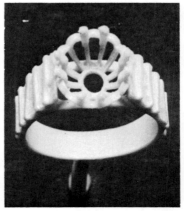

Fig. 82. Hard wax ring patterns for hand carving. Suitable for amateur use. Courtesy of Hoben Davis Ltd.

Sprueing and mounting techniques

Design of sprue formers

The basic function of a sprue former is to support the pattern, or patterns, during the investing process. When removed from the invested flask, the depression it has created will form an entry point for the casting metal to be directed into the mould cavity and subsequently will act as a 'feeder head' for the casting to draw upon as it contracts and solidifies.

During pressure casting, the hemispherical depression will also form the actual crucible in which the metal is melted.

Fig. 83. Typical sprue formers shown in section.

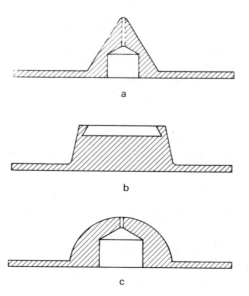

a

b

c

Fig. 83 illustrates the basic former shapes for centrifuge, vacuum and pressure casting. Type (a) may be used for single pattern mounting and a short piece of small diameter rod may be fixed in the hole to act as a reinforcing member. Alternatively, a cylinder of wax may be fixed to the top of the cone when using the tree spruieng technique for multiple castings (Fig. 84).

Type (b) is suitable for vacuum and centrifuge casting, especially where a number of patterns are to be mounted, as in 'button' or 'gang' spruieng. The undercut, which is filled with wax, is an added refinement which will give a good key to the wax coating upon which the patterns are sprued.

Type (c) is designed for pressure casting so that the casting grain in the hemispherical depression can be heated evenly and, when fully molten, will present the maximum surface area for air or steam pressure.

Here, a metal sprue pin is essential to support the pattern and to form the hole through which the molten metal will be forced into the mould cavity.

The diameter of the sprue pin is fairly critical; too small and the metal may 'freeze' or solidify before the mould is filled; too large and the molten metal may 'leak' in and again solidify before pressure from the caster can force it into all the fine detail. A diameter of 1.6 mm. (or $\frac{1}{16}''$) is about right and mild steel welding rod, cut to suitable length, is ideal for the purpose.

Sprue bases or formers are generally made from aluminium and some form of lip around the edge is useful where only one diameter of flask is to be used. An effective seal between sprue base and flask, to prevent seepage of the liquid investment, may be made with a fillet of wax or plasticine.

Rubber sprue bases, frequently favoured by trade casters, are made so that the flexible outer walls make a snug fit around the flask, thus eliminating the 'non-productive' time taken in sealing with wax or plasticine (see Fig. 85). Such sprue bases are often made by the trade caster using a metal die and a rubber vulcanising press.

A makeshift but equally effective former for pressure or centrifugal casting may be readily made from a suitably shaped piece of plasticine pressed firmly on to a piece of aluminium plate of sufficient area to contain the flask.

This particular method is suitable for single pattern and pattern clusters requiring a larger volume of casting metal.

Once the investment has set and the base plate

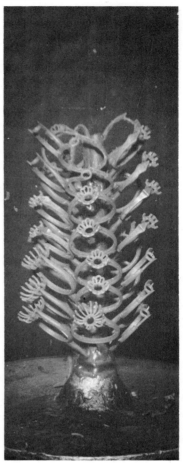

Fig. 84. An example of tree spruieng on a conical aluminium former.

66

has been separated, the plasticine may be removed with a modelling tool, and any investment flash trimmed away with a knife or carver.

Mounting the pattern

Perhaps the most important considerations here are the ensuring of an adequate supply of metal to all parts of the mould before the molten metal solidifies, and the provision of a suitable reservoir of metal which the casting can 'draw' upon as it solidifies.

. Lack of adequate sprueing can result in incomplete castings, shrinkage in the heavier sections and sometimes massive internal porosity within the body of the casting.

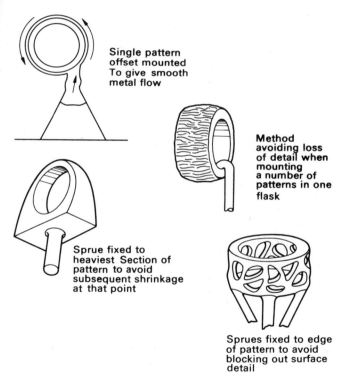

Single pattern
offset mounted
To give smooth
metal flow

Method
avoiding loss
of detail when
mounting
a number of
patterns in one
flask

Sprue fixed to
heaviest Section of
pattern to avoid
subsequent shrinkage
at that point

Sprues fixed to edge
of pattern to avoid
blocking out surface
detail

Fig. 86.

It is important, therefore, to ensure that sprues are as short as possible and are directed towards the heaviest section of the pattern.

Unfortunately, it is not possible to recommend any one method of sprueing as different patterns necessarily require different techniques.

Figs 86–9 illustrate some of the accepted methods

Fig. 87.

Smooth fillets ensuring
even metal flow

Rough incomplete joint-pattern
may break off or casting be
incomplete through premature
cooling of metal at this point

Thin sprues
to act as air
vents and feed
extremities
of the casting

Sprue too
long-metal
could freeze
before cavity is
filled

and the appended notes may give some guidance.

Wherever possible, smooth fillets should be formed at the joining points of pattern and sprue. Not only does this avoid sharp corners on the investment, which could get broken away and carried into the mould cavity by the metal flow, but the fillets also help to maintain fluidity of the molten metal by cutting down turbulence in the sprue cavities.

It is vital that the joints between pattern, sprue and base are firmly 'welded' together as it is most frustrating, when casting has been completed, to find that the pattern has come adrift before the investment has set and the whole exercise has been a waste of time.

A modelling tool heated over a spirit lamp or a modified electric soldering iron with an operating temperature of about 100°C, will fuse the wax components together most effectively.

The next stage is to coat the pattern with a thin film of wetting agent or 'de-bubbliser', such as 'WETTAX'. This, as mentioned earlier, will release

Fig. 88.

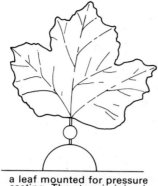

3 wedding bands made in one piece to take up minimum width in the flask. Parting-off may be done after casting

a leaf mounted for pressure casting. The stem is joined to the steel sprue by the wax sprue reservoir

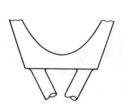

Sprues should be as thick as the heaviest section of the pattern

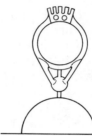

Spruing a delicate ring for pressure casting

the surface tension of the wax, allowing the liquid investment to make intimate contact with the fine surface detail of the pattern.

Fig. 89.

Pattern mounted at an angle to allow air bubbles in the investment to slide off easily. Note sprues fixed radially to ensure even flow of metal to all parts

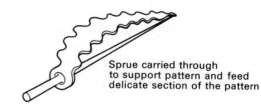

Sprue carried through to support pattern and feed delicate section of the pattern

It is important that this thin film is allowed to dry before investing takes place, otherwise it could cause air bubbles to remain in contact with the pattern.

Where the investment is to be de-aerated under vacuum, the use of the wetting agent may be omitted, though its use will not have any detrimental effect upon the resultant casting.

A suitably sized metal flask is fitted to the former and, where the latter also is made of metal, a seal is formed with plasticine or wax, as shown in Fig. 26.

Flasks may be made from any metal which will withstand the burnout temperatures. Although mild steel is sometimes used, the useful life is rather short due to repeated scaling from furnace heat.

Copper tube is often used by dental technicians when casting small inlays but the soft, malleable nature of the metal would easily allow distortion in larger flasks unless thicker gauge metal were used, and expense alone will generally rule out this possibility.

It is normally accepted that stainless steel is the most suitable material for flasks and although the initial cost may be relatively high, this is more than compensated by its longevity.

Depending upon overall size, suitable wall thicknesses range from 20 to 16 gauge and flasks as shown in Fig. 90 may be purchased from suppliers, cut from stock sized tube, or even rolled and welded.

Fig. 90. Stainless steel investment flasks. Courtesy of Hoben Davis Ltd.

Where high temperature investments are to be used for casting metals such as platinum and palladium, flasks made from inconel would be more suitable.

It would be difficult to recommend any definite size for a flask as the size and shape of patterns to be cast, and the capacity of the casting machine to be used, will obviously dictate MAXIMUM dimensions, though some casting machines will allow a range of different sized flasks to be used. However, some suggested sizes are listed in the materials and equipment section at the end of this chapter.

Where critical pattern dimensions are to be maintained in the final casting, it is advisable to fit an asbestos paper lining on the inside of the flask, as shown in Fig. 21. Experiments have shown* that flat plates of silver and gold alloys cast parallel to the axis of the flask lost up to 50% of the original pattern thickness when the lining was omitted. The reduction in thickness when a lining was used was in the order of 10%.

It is believed that the shrinkage is caused by the compression of the mould cavity, as the flask contracts more rapidly than the investment. The use of a lining not only offers a cushioning effect to mini-

* *A Decade of Jewellery Casting* by P. E. Gainsbury.

70

mise this contraction but also allows easier subsequent removal of the investment.

The lining is cut to size, allowing a small overlap where it joins, and should be shorter than the flask by about 6–12 mm at each end in order to allow the investment to bind to the inside wall of the flask.

Once the lining is fitted the whole flask is gently submerged in cold water to wet the asbestos and is then set aside to drain, prior to placing on the crucible former.

At this stage it is important not to touch the wet lining as pressure from the fingers is sure to compress the asbestos and impair its efficiency.

Where more than one flask is to be invested from a given mix it is a useful tip to have some sort of distinguishing mark on each flask so that a note may be taken of the type and amount of casting metal to be used in each case. Once invested they all look very much alike! The marks could take the form of file notches, as gummed labels and the like would be consumed during the burnout process.

Suggested materials and equipment

Basic
Crucible formers to suit type of casting machine
Metal sprue wires—about 1.6 mm ($\frac{1}{16}''$) dia.
Modelling wax
Sticky wax (model cement)
Plasticine
Wax sprue wires (i.e. round profile) of various diameters
Modelling tools
Thin lubricating oil
Spirit lamp
Wetting agent (de-bubbliser)
Soft artist's paint brush
Asbestos lining paper (approximately 1.6 mm thick)
Stainless steel casting flasks or cans. The following selection of sizes will cover most amateur requirements:

$$
\left.
\begin{array}{l}
57 \text{ mm (or } 2\frac{1}{4}'') \text{ high} \\
70 \text{ mm (or } 2\frac{3}{4}'') \quad ,, \\
90 \text{ mm (or } 3\frac{1}{2}'') \quad ,, \\
120 \text{ mm (or } 4\frac{3}{4}'') \quad ,,
\end{array}
\right\} \text{all by 63 mm (or } 2\frac{1}{2}'') \text{ 0/dia.}
$$

Note: These sizes are only approximate and small variations will be of no consequence. Where large

71

pattern clusters are to be cast, then diameters of 88.9 mm ($3\frac{1}{2}''$), or 101.6 mm ($4''$), would be more appropriate, though the actual size will naturally be determined by the capacity of the casting machine.

Adjunct
Modified electric soldering iron
Electronic wax knife with variable temperature control, complete with a range of shaped tips
Rubber sprue formers to suit diameter of flask

6   The investment process

The most widely used jewellery investment in the U.K. is gypsum based, with silica as the necessary refractory. Other modifying agents are added to control the degree of expansion and produce a reducing effect upon the molten metal when it enters the mould.

Whilst plaster based investments are quite satisfactory for silver and most gold alloys, they are unsuitable for some white golds and the high temperature alloys such as platinum and palladium. Here a high temperature silicate based investment must be used if mould fracture and sulphur contamination of the castings are to be avoided.

Dental investments have been widely used in the past and are still favoured by some craftsmen. These investments are necessarily subjected to a high quality control and are, in the main, considerably higher in price, so they are not of real interest to the beginner.

Due to the hygroscopic nature of the investment, care must be exercised in its storage. The powder should be contained in a strong polythene bag, preferably within an airtight container and stored in a dry atmosphere. Minimum exposure to air is recommended, particularly if casting is carried out at infrequent intervals.

Whichever type or brand of investment powder is used, the manufacturer's instructions concerning liquid/powder ratios should be followed closely if regular success in casting is to be assured.

Once the former—if made of metal—has been given a light film of oil to prevent the investment from adhering, the flask is fitted and a suitable seal formed. The necessary materials and equipment for investing may now be assembled.

In order to minimise wastage, it is a good tip to stamp each different sized flask with its maximum capacity of *mixed* investment. Not only will this economise on materials, it will avoid having hastily to mix extra investment when a flask is only partially filled from the first mix.

Fig. 22 illustrates the basic equipment required for mixing small batches of investment. Instead of using delicate weighing apparatus—which might easily be damaged in the workshop—a simple comparator can be made utilising empty plastic containers. Each container, which formerly held 2 oz of cream, when cleaned has a capacity of just over 100 grams.

Weights of 100 and 50 grams, or pieces of plasticine of the same weight may be placed in the opposite container to complete the balance. For all practical purposes this comparator will be quite suitable and sufficiently accurate for amateur use.

A graduated measuring cylinder, flexible plastic or rubber mixing bowl and curved spatula complete the kit.

The most common ratio of powder to water is in the order of 100 grams to 40 cm³ of water, so the next stage is to calculate the amount of water and powder required to fill the flask, and sufficient clean water for the whole mix is then put into the bowl.

The temperature of the water should be between 20°C and 22.5°C, bearing in mind that the warmer the water the more rapid the initial setting, thus cutting down the working time allowed to invest the pattern.

It is also important that the bowl and spatula be cleaned of all old investment prior to mixing a fresh batch, otherwise this again will accelerate the initial setting time.

The powder should be added to the water in, say 100 gram amounts and thoroughly mixed with the spatula, taking care to scrape off and mix in any lumps sticking to the side of the bowl. This should take about three minutes. When large amounts are to be mixed, a domestic food mixer having a stainless steel bowl may be used to great advantage.

One of the main problems facing the caster is the removal of air bubbles from the investment. Should these be left in contact with the pattern they will be converted into metal nodules on the final casting. This is shown clearly in Fig. 41, though in that instance their removal did not cause any problem.

74

Fig. 91. Vacuum mixing equipment. Courtesy of Metrodent Ltd.

On a more intricate casting, however, the removal of such nodules could be difficult and time consuming.

The mixed investment should now be vibrated, or subjected to vacuum for a short period of time, to eliminate the majority of the air trapped in the investment. Some workers, particularly in the dental profession, actually mix the investment under vacuum, using equipment similar to that shown in Fig. 91.

The next stage is to fill the flask; in Fig. 24 one can see the investment being poured alongside the pattern whilst the flask is vibrating on the table. The slurry, as the liquid investment is sometimes called, slowly rises up the flask, displacing the bulk of the air from the fine detail of the pattern. Unfortunately, no matter how much care has been taken with the spatulation and the initial air elimination, there are sure to be a few bubbles left in contact with the pattern and these must be removed.

The common methods of air elimination are described in the following paragraphs:

1. The simplest but least effective method, particularly where intricate patterns are involved, is by tapping the side of the flask with a metal rod. This has the effect of breaking the surface tension of the slurry, allowing the air to rise more readily.

2. The pattern is coated with a layer of liquid investment prior to preparation of the main mix. The investment is applied with a fine artist's brush and, once coated, is allowed partially to set before the flask is finally filled. This method is particularly applicable when intricate patterns or delicate 'natural' objects are to be cast.

3. The flask is held in both hands, and the edge of one hand is allowed to rest on an electro-magnetic vibrator. This method may be improved with the addition of an amplitude, or vibrating table, as shown in Fig. 92. Such a table may be readily constructed with basic equipment and will prove to be reasonably efficient for air removal.

The table, which rests on springs, could be surfaced with a 'wipe clean' material such as one of the melamines. A rim around all sides will prevent the flask from sliding off.

Not only does the use of a table leave both hands free; there is the added advantage that

Fig. 92. A simply constructed vibration or amplitude table.

more than one flask can be filled from one batch of investment.

4. The invested flask is subjected to a reduction in pressure equivalent to 29″ Hg by use of a vacuum pump (Fig. 93). Without a doubt this is the most efficient method of removing the free air, though the equipment can be expensive, particularly where a rotary high vacuum pump is used.

In the past, vacuum gauges were fitted to these machines and were often misleading, as the reading can be affected by atmospheric pressure. This sometimes led the operative to believe that his pump was failing.

The vacuum gauge indicates the degree of vacuum, in inches of mercury, that is being pumped by the machine and depends on atmospheric pressure of 14.7 p.s.i. which is sea-level pressure. If the surrounding altitude is increased, the ultimate vacuum reading will be proportionately that much lower. The following table indicates the equivalent pressures at various altitudes:

Fig. 93. Vacuuming a batch of invested flasks.

Altitude (in 1,000 feet)	Vacuum (in mm of Hg)	Pressure (in inches of Hg)	Pressure (in p.s.i.)
0	760	29.992	14.70
2.5	693.8	27.315	13.42
5	632.4	24.897	12.23
7.5	575.5	22.656	11.13
10	522.7	20.580	10.11

An easy rule of thumb method of checking the efficiency of the pump is to place a partially filled container of water in the vacuum chamber and apply the vacuum. The water should bubble violently within 60 secs* if the system is operating efficiently.

There are several types of vacuum investor currently available (Figs 94–6); some utilise a glass or reinforced plastic bell jar resting on a rubber or neoprene pad for the vacuum chamber, whilst more sophisticated models house the chamber, fitted with a transparent cover, within a cabinet.

Some—see Fig. 128—have the bell jar table mounted on 'joggle' springs similar to the vibrating table described in paragraph 3 (p. 75), the idea being that by thumping the corner of the table during

Fig. 94. Vacuum investing machine using a bell jar for the chamber. The front panel of the machine has been removed to reveal the vacuum pump.

* This time depends upon the type of pump used and the capacity of the vacuum chamber.

76

Fig. 95. Modern design in vacuum equipment. Courtesy of V. N. Barrett (Sales) Ltd.

Fig. 96. A vacuum chamber so designed that it can be mounted in a recessed work top, giving a near flush surface. The special loading tray will contain any spilled investment.

vacuuming, the flask will vibrate, allowing the air to escape more readily.

An interesting phenomena which occurs during vacuuming is the considerable increase in volume of the investment as the pressure drops. This, of course, could prove disastrous if the investment overflowed from the flask.

There are three widely used methods of counteracting this 'rise'. One involves the use of a proprietary 'de-foaming' agent; another is to have taller flasks than the pattern requires. The latter method is commonly used in the casting trade and is especially effective when the investment has been vacuumed prior to flasking. Only sufficient investment to cover the pattern adequately is poured into the flask and the remainder of the space is used to contain the rise.

The third, and most popular method open to the amateur, is to fit a stout paper or rubber collar around the top of the flask, so that it extends at least half the height of the flask above the rim. This collar is removed after the investment has set.

Vacuum investing machines, as mentioned earlier, are expensive pieces of machinery unless one is fortunate enough to obtain one second hand. However, a simple and reasonably efficient one may be made quite easily.

The model involves the use of a water jet pump as illustrated in Fig. 97, a stout airtight case with a transparent 'window', and a length of rubber vacuum tubing.

A chamber of approximately 130 mm³ will contain a 300-gram capacity flask comfortably and tests have shown that with a mains water pressure 50 lb/in², a vacuum gauge reading of 29 inches of mercury can be achieved well within one minute. The water consumption is in the region of 9 litres/min (2 gallons/min).

With some degree of improvisation the construction of this piece of equipment is well within the scope of the amateur. Fig. 98 shows the chamber resting on a neoprene surfaced wooden pad, though a more sophisticated dual purpose machine is featured in Fig. 99.

Once the flask has been subjected to vacuum for about 1 minute, the vacuum is released and the flask removed from the chamber. A couple of gentle taps on the side of the flask will help to settle the slurry,

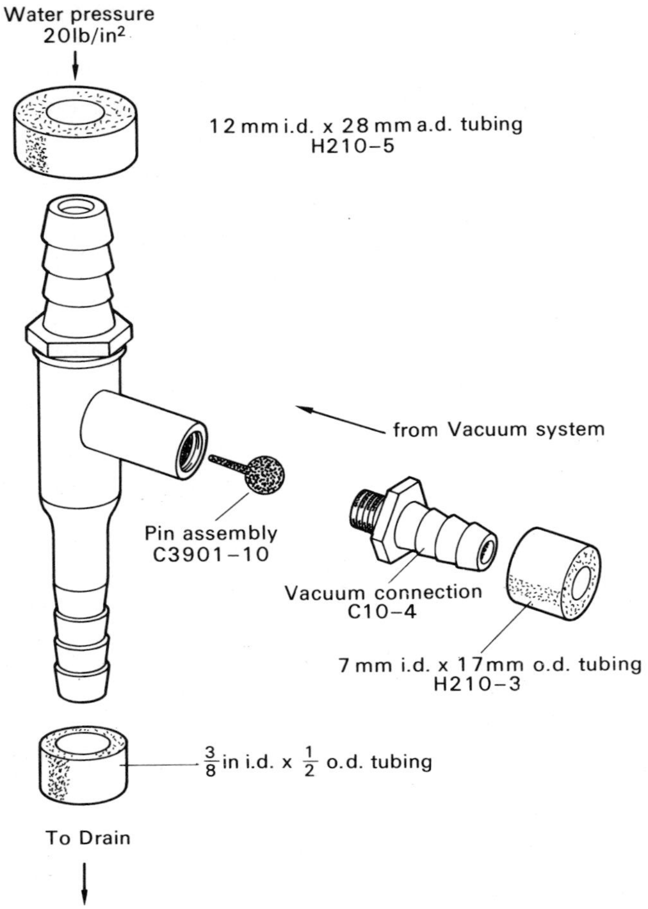

Water pressure
20lb/in^2

12 mm i.d. x 28 mm a.d. tubing
H210–5

from Vacuum system

Pin assembly
C3901–10

Vacuum connection
C10–4

7 mm i.d. x 17mm o.d. tubing
H210–3

$\frac{3}{8}$ in i.d. x $\frac{1}{2}$ o.d. tubing

To Drain

Fig. 97. A simple water jet pump.
Courtesy of Edwards Vacuum
Components Ltd.

and any topping up to cover the pattern to the required depth may now be done, with the surplus from the mix.

The flask is now put to one side whilst the tools and equipment are cleaned of all wet investment.

Note: The working time of 'Investrite' jewellery investment and Kerr's SATINCAST 20 (two of the commonly used investments) is 8–9 minutes and the manufacturers recommend that this time is fully used up. If it is found that the mixing and vacuuming can be completed in a shorter period then the mixing time should be increased; otherwise separation or stringing of the constituents may occur.

On no account should the vacuuming time be increased, as the slurry will stiffen when too much water is 'boiled' away and may not go back in intimate contact with the pattern.

78

Fig. 98. Test rig for a vacuum
investing chamber. The vacuum
gauge, inserted only for purposes
of illustration, shows a reading of
29.5" of mercury, which was
achieved in 35 seconds. The
vacuum release screw, not shown
in the illustration, is in the side
opposite to the air outlet.

Fig. 99. Combined vacuum and
investing equipment utilising a
water jet pump. Courtesy of
Hoben Davis Ltd.

Fig. 100. Balancing a horizontal
caster.

When 45–60 minutes have elapsed (longer when
a large flask is used) the sprue base, or former,
should be removed. Where a metal sprue pin has
been used to support the pattern it should be removed first.

It is wise to hold the flask, sprue hole down, when
removing the base in order to minimise the chance
of any small particles of dried investment entering
the mould cavity and possibly being included in the
casting.

Any surplus investment is now trimmed off with
a sharp knife or an old coarse file—one of the
'dreadnought' cut is ideal for the purpose—to ensure
even seating for both pressure and centrifuge casting
techniques.

Where a vertical centrifuge machine is to be used
for casting, the flask should be set in the machine,
the correct quantity of casting grain put in the
crucible and the balance weights adjusted until the
arm is stationary in the horizontal position. This is
shown in Fig. 29.

The method differs slightly when a horizontal
spring driven caster is used. Fig. 100 illustrates a
matchstick wedged between the straight arm and the
secondary arm. The centre pivot nut is loosened until
the arm rocks slightly on the pivot point and the
counter balance weights are adjusted until the arm
will rest in the horizontal position. Balance is
achieved when a slight tap on either end will cause
it to tilt in that direction. The centre nut is then
tightened, the matchstick removed and the short, or
secondary arm, returned to its cranked position.

The setting of the investment is the last step before
burnout, for during the hour or so it is left after the
air has been removed setting expansion takes place,
This expansion contributes substantially to the total
expansion of the mould cavity required to compensate for the expansion and eventual contraction of
the molten casting metal.

The contraction rate of gold alloys is in the order
of 1.25 to 1.5% and of this total amount the setting
expansion may contribute as much as a fifth.

Whilst the setting expansion takes place the mould
also gains in strength. Many casting failures, where
the fired investment in the mould cavity has broken
down under impact from the injected metal, may be
attributed to an insufficient time allowed for setting.

The next step in the process is the wax elimination

and burnout of the invested flask. Sometimes it may not be convenient for the investing and burnout stages to be consecutive and opinions vary over the method of storage prior to burnout.

Some authorities favour wrapping the flask in a damp rag and storing it in an airtight container until burnout is convenient. Others just leave the flask standing, sprue hole down, in some convenient place where it will not be disturbed.

Successful casts have been obtained using either method where periods as long as a week have elapsed between investing and burnout. With the latter method, however, it would be wise to line the inside of the flask with damp asbestos paper prior to investing, in order to compensate for the thermal expansion of the investment during burnout.

Suggested materials and equipment

Basic
Investment powder
Flexible mixing bowl
Curved mixing spatula
Investment proportioner, or accurate weighing scales
Graduated glass or plastic cylinder, 100 cm³ minimum capacity
Soft artist's brush
Electro-magnetic vibrator
Plasticine

Adjunct
Mechanical investment mixer
Vacuum machine
De-foaming agent

Plate I The finished ring.

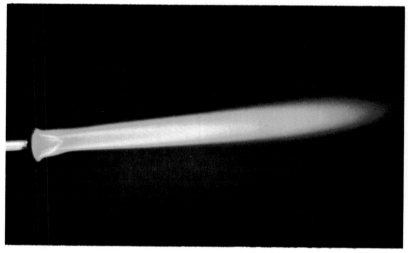

Plate II The reducing area is between the tips of the inner and outer cones.

Plate III A small selection of faceted synthetic gemstones. Courtesy of D. Swarovski & Co. International (U.K.) Ltd.

Plate IV — above and right — A selection of tumble-polished gemstones from Hillside Gems Ltd.

Left: Gold dress rings from the Richard Ogden Collection.

Below: Pendant necklace in gold with black and grey pearls. By Philippe Pateck. From the Goldsmiths collection.

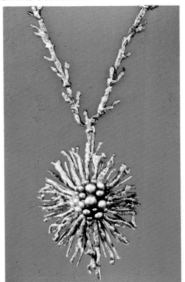

Below: Bracelet in oxidised silver with quartz. By J. Tharrats. From the Goldsmiths' collection.

Top, left and right: Castings of simple natural subjects.

Above: Textured 18ct gold necklace set with moonstones and diamonds. By Gilian Packard.

Right: Set of jewellery made in 18ct yellow gold flame shape gold and set with turquoise, sapphires and diamonds. By John Donald.

Top left: 18ct gold bracelet set with emeralds. By Gilian Packard.

Middle left: 'Egyptian influence'. Promotional jewellery for the Tutankhamun exhibition by Thomas Frattorini Ltd. Courtesy of Times Newspapers Ltd.

Bottom left: Sterling silver brooch set with amethyst spheres. By Gilian Packard.

Top right: Necklet made in 18ct yellow wavy rod gold and set with stick coral and pave set diamonds. By John Donald.

Bottom right: Sterling silver cuff links set with amethyst spheres. By Gilian Packard.

7 Wax elimination and burnout

All the painstaking work that has been accomplished up to this stage will be of no avail if correct burnout procedures are not observed.

In their excellent publication *A Handbook of Dental Procedures*, J. F. Jelenko & Co. Inc. summarise the aims of a successful burnout as follows:

1. To eliminate all moisture from the invested ring (flask).
2. To eliminate from the mould cavity the volatile portions of the wax or plastic used to form the pattern.
3. To eliminate from the mould cavity, or from the interstices of the surrounding investment, the carbon residue remaining from the wax.
4. To raise the temperature of the mould to the proper point to receive the molten metal, when the cast is made.
5. Through temperature rise, to produce the necessary expansion in the investment to compensate for the shrinkage of the gold after casting and cooling.
6. Through proper control of temperature, to prevent damage to the investment through overheating, which can lead to coarse grain and weakness in the gold; rough surfaces resulting from breakdown of cavity walls; and sulphur contaminated gold.

Although a variety of makeshift furnaces can be utilised for burnout purposes* and, with some experimentation, a reasonably high degree of success may

* See *Lost Wax Air Pressure Casting* published in the Griffin Technical Studies series.

81

be achieved, it is far better to obtain a furnace specifically designed for the job.

The conditions within a furnace muffle should be of an oxidising nature if all the carbon residues are to be eliminated.

One only has to observe a piece of smouldering charcoal when a jet of air is directed on to it. The charcoal will immediately glow and begin to be consumed. With the air blast removed, it will cease to glow and quickly revert to its original state, and, at the expense of considerable ash residue, may take several hours to be consumed.

Both gas-fired and electric muffle furnaces are used successfully for burnout. The gas type, although presenting certain problems with ventilation and flame adjustment to give the necessary oxidising conditions, is generally more favoured by trade casters (Figs 101 and 102).

For the amateur, the electric muffle furnace seems to be the obvious choice, and a number of eminently suitable models are currently available (Figs 103–5). For this reason the subsequent burnout procedures will be described using this type of furnace.

As mentioned earlier, the furnace chosen—and it pays to 'shop around'—should be designed specifically for wax elimination, if one is to avoid purchasing a de-waxing oven. These ovens are used frequently in the engineering precision casting industry for reclaiming wax. However, as wax usage in jewellery casting is relatively small, for the purposes of this publication they can be discounted.

The furnace muffle (chamber) should be large enough to take one, or more, flasks and refractory trays necessary to collect the wax and protect the furnace floor. A tube, or vent, from the rear of the muffle will exhaust the wax fumes and help maintain an oxidising atmosphere.

Unless burnout is going to be a hit or miss exercise, furnace controls are obviously necessary. These could include a temperature indicator, an energy controller to regulate the rate of rise in temperature and, if possible, a solid state controller.

The controller will allow the craftsman to dial the temperature desired and leave the furnace to do the rest, the rate of climb being dictated by the energy control.

This system of control has the obvious advantage that the operator does not have frequently to inspect

Fig. 101. A free standing gas-fired burnout furnace. Courtesy of Kasenit Ltd.

Fig. 102. An industrial gas-fired burnout furnace.

Figs 103 & 104. Electric burnout furnaces of modern design. Courtesy of Metrodent Ltd.

the temperature indicator through the period of burn-out. A furnace featuring all these controls is illustrated in Fig. 105.

Electric enamelling kilns have been used for burnout purposes but those having exposed elements should be avoided, as the useful life of the elements can be reduced considerably by water vapour and wax fumes.

Opinions vary as to whether the flask should be put in a cold or pre-heated furnace, though the author favours the use of a muffle pre-heated to a temperature of about 150°C.

The flask is placed, sprue hole downwards, on a small stainless steel or refractory tray within the furnace muffle. When only one flask is to be burnt out a position near the back of the furnace will ensure more even heating.

The use of the tray not only elevates the flask to allow the liquid wax to escape more readily, it also protects the elements under the muffle lining from wax contamination.

In order to melt out, or 'lose' the wax, a period of time must be allowed to elapse for the investment to reach the same temperature as the muffle. No matter how much higher the temperature of the muffle may be, the temperature of the mould will not rise much above 100°C until all the moisture has been eliminated.

It could, with certain exceptions, be most unwise to place a flask containing a plaster/silica investment in a muffle of elevated temperature, as the trapped moisture may be converted explosively to steam, with rather dramatic results!

Depending upon the size of the flask, an hour is usually sufficient time to eliminate all the volatile wax. Crackling, sizzling or 'frying' sounds are characteristic during this stage and indicate that the wax has not yet been 'lost'.

With the wax eliminated, the tray may now be removed and the flask laid on its side with the sprue hole facing the furnace door. This position is advised if oxidising conditions within the mould are to be achieved, and localised 'hot spots' avoided on the muffle floor.

During burnout the surface of the investment will show considerable discolouration. One indication of the fired state of the investment is the colour change from black, through grey/yellow, to bone white when the investment is completely free of moisture.

The temperature of the muffle is raised gradually to a maximum of between 700°C and 730°C, and because of the time lag between flask and furnace temperatures, the flask is allowed to 'heat soak' at the upper temperature for a period of at least 1½ hours.

Visual inspection of the sprue hole when the mould nears this temperature will show the cavity glowing dull red. One must not now assume that the flask is ready to cast. Remembering the example of the charcoal, it is only at this stage that the carbon residues will burn away.

Fig. 106 illustrates the chart issued by the manufacturers of 'Investrite' jewellery casting investment, which indicates conservative times for burnout.

It should be borne in mind that the times indicated would probably be for the larger flasks used in the jewellery casting trade and one should safely be able to shorten the heat soak and temperature reduction periods when using smaller flasks.

When burning out patterns constructed from natural or 'found' objects such as leaves, seed pods, pieces of bark, etc., no exact timings can be recommended. Although muffle temperatures will be the same as for invested wax patterns, the duration of the heat soak period must necessarily be longer if hard or woody specimens have been used. Patience and visual inspection of the mould as viewed through the sprue hole are the only ways to tell if the specimen has been consumed.

When burnout has been completed it may help to remove any remaining ash residue if the side of the flask is tapped gently with a metal rod whilst it is being transferred to the casting area. Make sure that the sprue hole is facing down when you do this; otherwise an incomplete casting may result.

Leading authorities advise that following the 'heat-soak' period, the flask temperature should be lowered, the actual amount depending on the design of the pattern and the type of casting metal. This is sound advice, as too hot a mould when the metal is injected can give rise to porosity and grain growth in the final casting.

The following burnout cycles are recommended by the Kerr Manufacturing Co. for their Satincast 20 and Brilcast investments and will serve as a useful guide for similar investments.

Fig. 105. A moderately priced muffle furnace designed specifically for wax burnout, though equally effective for enamelling and heat treatment. Courtesy of British Ceramic Service Co. Ltd.

5-hour cycle	8-hour cycle	12-hour cycle
For flasks up to 2½″ × 2½″ Pre-heat furnace to 150°C 1 hour—150°C 1 hour—371°C 2 hours—732°C	For flasks up to 3½″ × 4″ Pre-heat furnace to 150°C 2 hours—150°C 2 hours—371°C 3 hours—732°C	For flasks up to 4″ × 8″ Pre-heat furnace to 150°C 2 hours—150°C 2 hours—371°C 2 hours—482°C 4 hours—732°C
1 hour reduce flask to proper casting temperature*	1 hour reduce flask to proper casting temperature*	2 hours reduce flask to proper casting temperature*

* Mould temperatures for ladies' rings or items of lacy or intricate design should be 537°C–593°C. Mould temperatures for men's rings and items of relatively heavier design should be 371°C–482°C.

Fig. 106. Burnout cycle for gold and silver using Investrite.

These figures are straight conversions from the Fahrenheit scale and may safely be rounded up or down for the sake of convenience, i.e. 732°C may be read as 730°C.

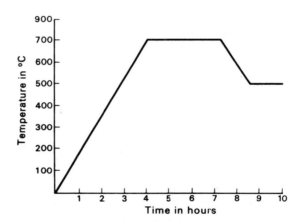

In his paper, *Investment Casting of Precious Metals*, Peter Gainsbury quotes mould temperatures used generally when casting precious metals as:

9 carat Gold alloys	500°C–600°C
Silver alloys 18 carat Gold alloys	600°C–700°C
White Gold alloys Palladium alloys Platinum alloys	700°C–800°C

Three important rules should be remembered if successful burnouts are to be achieved.

1. Once started, the burnout cycle should be continued or the flask discarded. It could be most unwise to re-heat a partially fired flask from a previous day. With the removal of moisture by the initial firing, the subsequent expansion occurring during the second burnout would most probably cause serious cracking of the mould.
2. The temperature rise up to 500°C should be gradual and controlled, though once the flask has reached this temperature a more rapid elevation may be allowed without harming the investment.
3. Except when using high temperature investments, such as Kerr's Ferrolite/Platinite, the flask temperature should not be allowed to exceed 780°C. After this temperature, the plaster/silica investments will break down progressively, with the possibility of sulphur contamination and cracking of the mould, both leading to serious defects in the resulting casting.

The beginner will find it a useful exercise to prepare flasks and experiment with burnout cycles, similar to those already mentioned, to suit his own particular needs.

Instead of completing the exercise by casting and quenching, each flask should be allowed to cool and, when the investment is broken out, the mould cavity may be examined for traces of carbon residue.

It may well be argued, and with some justification, that the usual burnout period is rather a lengthy affair and could disrupt normal routines in schools and colleges, etc. There is, however, another burnout cycle open to the amateur, which utilises the 'Baker Quick Heat Process', using STERLING Scientific Investment.

Although primarily intended for dental technicians, this plaster/silica investment may be used for making all types of moulds for casting precious metal alloys and exhibits considerable inherent strength and resistance to fracture.

With the 'Quick Heat' method the flask is allowed to bench set for only 20 minutes. During the next 10 minutes the sprue base, or former, is removed, the base of the mould roughened to facilitate easy escape of gases during casting and any balancing of the casting machine completed.

At the end of this period—30 minutes from the commencement of the mix, the flask is immersed in

86

cold water for 10 minutes. This is to allow for sufficient hygroscopic expansion to take place to compensate for the reduced time allowed for setting expansion.

These processes will take a total time of 40 minutes and the flask may now be placed in the furnace which has been pre-heated to 700°C–750°C. The temperature is maintained for approximately 45 minutes to 1 hour, depending upon the size of the flask, and at the end of this period casting may take place.

The manufacturers claim that, provided the flask has been lined with damp asbestos to allow hygroscopic expansion to take place, and the correct water/powder ratio maintained, with an efficient furnace, even the largest flasks should be ready for casting in 1 hour and 40 minutes from the commencement of the mix.

This investment, because of its considerable strength, does have one minor drawback. Unlike the common jewellery investments which disintegrate when the hot flask is quenched in water, Sterling often remains quite stiff. Some care therefore, has to be exercised when gouging out the investment if damage to intricate castings is to be avoided.

Although more expensive than jewellery investment, where time is at a premium, Sterling Scientific Investment offers possibilities for the amateur caster.

Suggested materials and equipment

Basic
Gas or electric muffle furnace capable of reaching and maintaining 800°C
Flask tongs
Stainless steel or ceramic trays for wax collection
Asbestos gloves

Adjunct
  Programme controlled gas or electric burnout furnace
or Electric muffle furnace with solid state controller

8 Casting

While the burnout of the flask is nearing completion the equipment and materials required for casting should be assembled.

Metal may be in the form of casting grain, which generally resemble small balls or nodules, or clean oxide free scrap may be used. A word of warning though: on no account should 'once cast' metal still bearing traces of investment be used. Under the fierce localised heat of the melting torch the investment will break down and release sulphurous compounds into the molten metal. Any traces of silver solder, too, should be removed as they will not mix homogeneously with the molten metal and will cause localised hard and soft spots. If 'once cast' metal is to be used it is advisable to include at least 50% new metal in the melt in order to maintain a reasonable balance of alloying constituents.

Old sprue buttons should be cut into smaller pieces and re-melted on a charcoal block. A shallow depression scraped in the surface of the block with an old knife will retain the metal when it becomes fluid.

Surface oxides may be kept to the minimum if a pinch of flux is sprinkled on the metal and melting is carried out in the reducing portion of the flame (see Plate II). Once the metal is fully molten and appearing to spin, the flame is removed and the metal allowed to solidify. Immersion in a hot pickle solution and a final rinse will remove any remaining oxides.

Where a spring driven or motorised centrifugal casting machine is to be used, the casting crucible should be examined for defects, such as cracks, exit hole partially blocked by old casting metal, and heavy oxide residues from previous melts.

A flask contaminated from oxide and flux residues

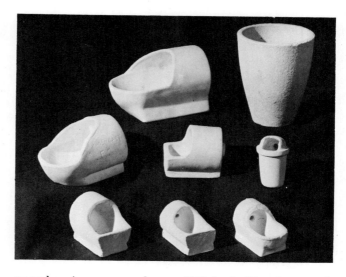

Fig. 108. A lightweight torch suitable for delicate soldering operations using natural gas with mouth-blown air or foot bellows.

may be given a new lease of life by boiling in a weak solution of sulphuric acid and water. A subsequent soaking in a neutralising solution of sodium bicarbonate, and a final rinse in cold water is necessary to prevent the crucible from disintegrating from acid attack.

A natural drying out period should be allowed, however, as any attempt to accelerate drying by torch or furnace heating will, almost certainly, cause the crucible to crack.

Some authorities advocate not only fluxing the inside of the crucible with borax but also lining it with asbestos for each cast. However, crucibles are relatively cheap and where one uses a fairly limited range of casting metals, a separate fluxed crucible may be used for each type of metal.

Although the modern crucible will withstand the sudden thermal shock from the heating torch, it is by no means a bad plan to pre-heat it in the furnace along with the flask. The pre-heating will certainly cut down the amount of time required to bring the metal to melting temperature and help to maintain its fluidity when the casting machine is set in motion.

Heating equipment

With the exception of the high frequency casting machine, it is necessary to use a torch for heating and melting the casting metal. For comparatively small melts of silver and most golds, an ordinary brazing or soldering torch, relying upon a compressed air/gas mixture is generally satisfactory (Figs 108–10).

89

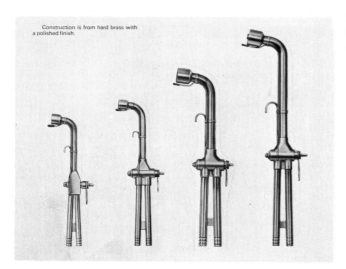

Construction is from hard brass with a polished finish.

Fig. 109. A range of torches for natural gas and air. Courtesy of Adaptogas Ltd.

Fig. 110. A lightweight brazing torch suitable for melting gold and silver. Courtesy of Flamefast Engineering Ltd.

Some trade casters favour the use of oxyacetylene, oxy-coal gas, or oxy-propane, and because these mixtures allow very rapid heating, the amateur should explore their use with considerable care if oxidising flames and over-heated metal are to be avoided.

With the appropriate nozzle, a propane torch also may be used successfully for small melts of gold and silver, this method having the advantage of portability over the static compressor units (Fig. 111).

Flame adjustment is vital during the melting stage and every endeavour should be made to maintain a reducing atmosphere over the melting metal in order to prevent the formation of harmful oxides.

Plate II shows a typical reducing flame, where the outer cone is fully combusted gas, whilst the inner cone comprises partially burned gas and is, therefore, of lower temperature. Melting should be done in the part of the flame between the tips of the inner and outer cones, where there is a slight excess of gas over air.

Fig. 111. A Sievert propane torch with an assortment of nozzles.

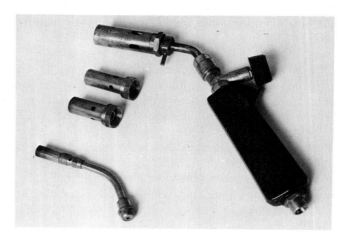

A certain amount of practice is necessary in order to achieve a reducing flame. Apart from careful adjustment of the controls, the torch should be moved gently towards and away from the melt, whilst the effect of the flame on the metal is closely watched. A dull, drossy surface on the silver or gold indicates an oxidising atmosphere, whilst a bright, shiny surface means that the correct reducing flame is being used.*

A pinch of flux will help to remove more persistent oxides and improve the fluidity of the metal just before casting. Suitable fluxes are boric acid, fused borax, or one of the proprietary products supplied specifically for this purpose.

Although many craftsmen favour the use of ordinary borax for fluxing purposes when hard soldering, its use is not recommended for casting purposes, as borax is hygroscopic and with the release of water vapour this is a strong possibility of gas porosity in the final casting.

The high temperature metals of the platinum group require an oxidising flame during melting. They are generally of a sluggish nature when fully molten, and where a spring-driven machine is being used a stronger spring setting may be necessary.

* Reference should be made to the bullion dealers' charts and tables before heating any of the various carat golds which are currently available. One dealer lists no less than 29 types of gold, though not all are suitable for casting.

To sum up then: The metal should be melted as quickly as possible, without overheating, using a reducing flame. Flux if necessary and when the metal is shining, fully molten, and spinning in the crucible, cast.

Choice of a casting machine

The choice of a casting machine obviously depends upon the amount and nature of the work to be done. Where funds are limited, or where relatively small items such as rings, brooches and pendants are to be cast on a one or two off basis, pressure casting is the obvious choice.

This method relies upon the pressure generated by steam or escaping air to force the molten metal into the mould cavity.

STEAM OR SOLBRIG METHOD

Favoured by countless dental technicians prior to the advent of a reliable centrifuge caster, this method, where sprueing and mounting of the pattern is identical with that of air pressure casting, relies upon the build up of steam pressure when a wet asbestos pad is brought in contact with the hot flask containing the molten metal in the crucible depression.

There have even been reports of successful casts made when a lump of wet clay is pressed on to the top of the flask, though the author regards this method with some misgivings, especially where personal safety is concerned.

A steam pressure caster is very simple to construct. In its simplest form a tin lid fitted with a wooden handle and a renewable asbestos wet pad (Fig. 112) will work reasonably well, though the system illustrated in Fig. 113 is more common.

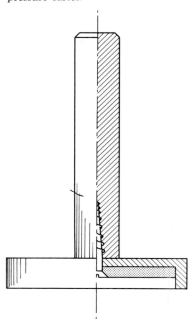

Fig. 112. A simple portable steam pressure caster.

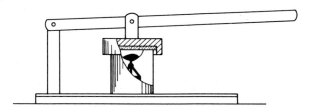

Fig. 113. Schematic diagram of a conventional Solbrig steam pressure caster.

Although this method has its faithful following, the author has never achieved a success rate higher than 70% and for this reason, of the two systems, personally favours air pressure casting.

92

Fig. 114. An air pressure casting
kit based on an original design
by the author. Courtesy of Griffin
and George Ltd.

Fig. 115. A casting of two silver
rings with the residual sprue
button of total weight 56.6 grams.
Student work using an air
pressure caster.

AIR PRESSURE CASTING

Already described in Chapter 2, this method is simple, clean, safe, and efficient. The pressure casting kit featured in Fig. 114 is currently available and requires only a footpump for priming. The pressure release valve on the top of the caster is a built in safety device to prevent the latter from being over pressurised, and when fully primed the valve will make a slight hissing sound as the surplus air is released.

From time to time it is wise to check the rubber sealing washers for any escape of air when the caster is pressurised. This is best done by painting a solution of soap and water around the junctions between the body and casting head and between the pressure release valve and the top of the caster. Should air bubbles appear, then the respective parts may be separated and a thin layer of vaseline smeared on both sides of the washers. When re-checking after assembly and re-pressurising, provided that the seals have not deteriorated with age, it will be found that this method usually remedies the fault.

Students working under the direction of the author have not only successfully cast such delicate replicas as leaves and insects but also relatively bulky castings as illustrated in Fig. 115, where two rings were sprued on to the sprue-reservoir.

CENTRIFUGAL CASTING

This method is by no means a modern innovation. Early craftsmen often employed an incredible variety of 'Heath Robinson' systems in their endeavour to obtain sound, dense castings.

Some dental technicians still use the 'bell and chain' method, though for safety reasons this procedure should be avoided in schools or work rooms where space is limited. Fig. 116 illustrates sketches of some of the early centrifuge casters.

Centrifuge casting machines may be classified broadly as vertical or horizontal. A machine whose casting arm rotates on a vertical spindle is known as a *horizontal* machine, whilst one whose arm rotates on a horizontal spindle is termed a *vertical* caster.

Almost without exception, vertical casting machines are spring driven, whilst horizontal casters may be spring driven or motorised. Some of the motorised machines can be expensive and

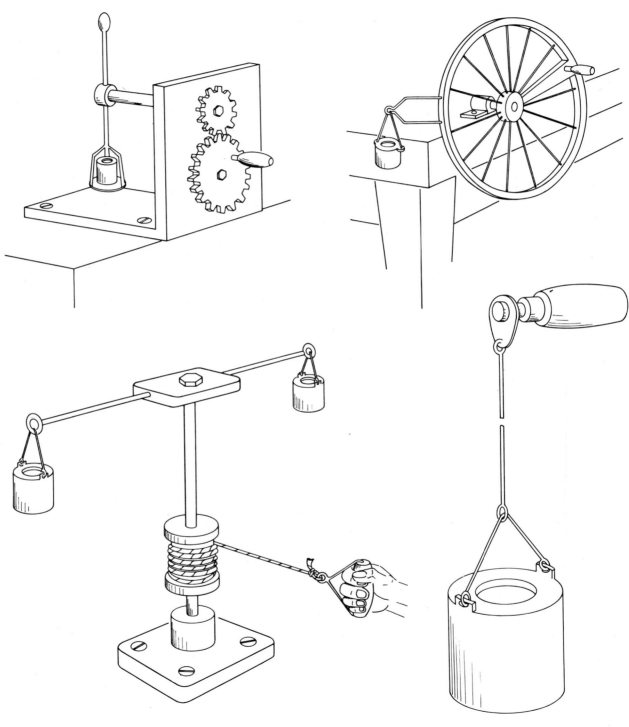

Fig. 116. Some early patented centrifuge casting machines.

94

Fig. 117. A vertical caster mounted on a work bench.

Fig. 118. Melting a charge in a Hooker caster. Courtesy of W. J. Hooker Ltd.

extremely sophisticated, their cost sometimes running into thousands of pounds. Such machines, in the main, would be confined to trade casters with a high productivity level.

The vertical caster

Fig. 117 shows a typical machine bolted to the top of a work bench so that waxing, investing and casting need occupy only minimum space. Equipment and materials may be stored below, whilst the castors will enable the bench to be moved easily to the heat source, where the latter is static. This type of machine relies upon tension springs for its motive force, different tensions being regulated by a trigger or rachet mechanism when a stronger 'throw' is required. The caster is supplied with a protective cowl and must be bolted securely to a firm surface. Depending upon the machine and type of crucible used, casting capacities

95

Fig. 119. Melting a charge of casting grain in a Kerr horizontal caster.

vary between 77.7 grams (2½ oz) and 466.5 grams (15 oz).

Along with pressure casting, the use of this machine method has already been described in Chapter 2, though specifications and explicit operating instructions may be obtained from the manufacturer.

Horizontal casters (spring driven)

Unlike the vertical caster, this type of machine relies upon an extremely strong coil spring, the strength of throw being governed by the number of turns given when the machine is 'wound up'.

Because the casting arm revolves in a horizontal plane, unlike the vertical caster, there is a tendency for the molten metal to be thrown out of the side of the crucible with the initial surge of centrifugal force. For this reason most machines have the arm supporting the cradle and flask hinged on a pivot. Thus, with the arm cranked at right angles towards the counterbalance assembly, the initial surge occurring when the stop rod is released is taken up by the crank mechanism.

The caster is bolted to a firm surface and, because of safety and possible fire hazards, should be contained within a sturdy metal bowl or a metal frame lined with an insulating material such as sindanyo board.

Similar to the vertical models, the horizontal caster should be balanced with its flask and crucible

96

containing the casting grain, prior to burnout, though the method of balancing, as already noted, is slightly different.

To wind the machine, the counterweight end of the arm should be grasped firmly and given three complete turns in a clockwise direction. With the spring wound, the stop rod is pulled up from the top of the spring housing and the casting arm allowed to lock against it.

The primed machine, with its casting arm cranked towards the counterbalance assembly, is illustrated in Fig. 119, where the casting grain is being heated by a portable propane torch.

When the metal is ready for casting the counterbalance arm should be moved slightly away from the stop rod, allowing the latter to drop back into the spring housing. Simultaneously, the arm should be released and the torch lifted up and away from the crucible. On no account should any attempt be made

Fig. 120. An American horizontal caster complete with kit. Courtesy of Kerr Manufacturing Co.

Fig. 121. A powerful continental horizontal caster. Courtesy of Metrodent Ltd.

to stop or shorten the duration of spin, as this is not only dangerous but it could affect the density of the casting. Figs 120 and 121 show two available machines.

Depending upon the model, these casters may have casting capacities between 77.7 grams (2½ oz) and 217.7 grams (7 oz) of gold. Machines with larger casting potential are available but vary slightly in design from those already discussed and would fall outside the scope of the amateur.

Motorised machines
These machines are primarily intended for production

97

workshops and generally have considerably more casting potential than their spring-driven counterparts. The centrifuge arm is driven by a powerful electric motor, usually clutch controlled. In some models provision is made for the melting torch to be held in a fixed position, so that the operator is free to use both hands during melting and casting operations. Figs 122 and 123 illustrate typical motorised casting machines.

HIGH FREQUENCY CASTING MACHINES

A really first class exponent of investment casting can achieve remarkable quality and variety with any equipment of a reasonable degree of performance. However, all the careful preparation in making the wax patterns, mounting, investing, and burnout may be to no avail if he misjudges the casting temperature of his metal and, through overheating, boils off some of the basic elements of the alloy.

Casting skill or expertise is usually built up through the medium of sound teaching and a considerable backlog of experience.

The main advantage of the high frequency machine, which naturally will have to be set against the considerable outlay, is its ability to melt the alloy rapidly by means of a high frequency induction coil, which is electronically controlled by a high temperature pyrometer to prevent overheating, with the formation of harmful oxides kept to the absolute minimum.

Thus, apart from increased casting potential (some models being capable of casting up to 3 kg of 9 carat alloy), these machines are a serious attempt to de-skill the routine of repetitive work, which is the greater part of the trade caster's bread and butter.

As seen in Figs 125 and 126 the machine is floor standing and is enclosed in a sturdy metal cabinet, which houses the high frequency power-pack and the electric drive unit for the centrifuge. The latter turns at two speeds, selection being dependent upon the type of casting being performed.

The front of the cabinet carries the control console which incorporates the ammeter, the main on/off switch, a switch for energising the centrifuge drive motor, and the lever for raising and lowering the heating coil.

On top of the cabinet is a deep metal safety shield which surrounds the centrifuge arm. This arm carries

Fig. 122. A continental motorised casting machine. Courtesy of Metrodent Ltd.

Fig. 123. A typical motorised horizontal casting machine. Courtesy of Hoben Davis Ltd.

Fig. 124. The casting arm of the same machine.

98

Fig. 125. A modern high frequency casting machine clearly showing the thermocouple probe. Courtesy of Ferraris Engineering and Development Co.

Fig. 126. A charge being melted in the same machine.

Fig. 127. Graphite and ceramic crucibles for the high frequency machine. Courtesy of Ferraris Engineering and Development Co.

the invested flask and crucible which, for casting silver and low temperature gold alloys, is made of graphite. This material enhances the efficiency of the high frequency heating effect and also helps to inhibit oxidisation. For platinum and high melting point alloys special ceramic crucibles are used (Fig. 127).

The sequence of operations is as follows: The heating coil is raised into position around the crucible using the lever provided, the power switched on and fusion of the metal in the crucible rapidly achieved. The temperature of the molten metal is monitored and controlled by means of the thermocouple which, in the meantime, has been lowered into the crucible. When the optimum casting temperature has been reached the thermocouple is raised clear, the heating coil lowered and the centrifuge operated by means of the switch. Electrical safety interlocks are included in the circuit to prevent the motor being started when the coil is in the raised position.

Although well beyond the pocket of the enthusiastic amateur, it is significant to note that this type of machine represents a technological advance beyond the normal motorised machine, comparable with that of a spring driven centrifuge over the traditional 'bell and chain' method.

VACUUM CASTING

Although widely used in America, this method of casting is relatively new to jewellery casting in the U.K. In the real sense of the word it is not a true

method of vacuum casting and could best be described as vacuum assisted gravity casting.

In centrifugal casting, the centrifugal force enables the molten metal to eliminate the air from the mould cavity through the porous investment, via the back of the flask. With the vacuum assist method, the pull of the vacuum is used to reduce pressure within the mould cavity, with atmospheric pressure acting to press the metal down into the mould.

The spruing technique, as described by the manufacturers of the American motorised machine illustrated in Fig. 128 is slightly different, in that the the pattern should be placed as low down in the flask as possible, yet leaving sufficient investment between the casting table and the pattern so that the vacuum will not cause a rupture in the investment at this point. They also recommend that wax sprues should be about 25% greater in diameter than those used for centrifuge casting.

Irrespective of which type of machine is used—motorised or water pump, it is essential to check that any irregularities are removed, both from flask rim and investment, so that an effective seal may be obtained with the heat-resistant seal on the casting platform.

Fig. 128. Combined investing and vacuum casting machine utilising a rotary pump. Courtesy of South West Smelting Co.

Casting with the motorised unit

The casting grain should be preheated in a small graphite crucible fitted with a pouring handle, just prior to the removal of the flask from the burnout furnace.

The flask is placed over the air extraction hole on the casting platform and the vacuum unit activated. With a good seal obtained (and it is wise to check this by pressing against the side of the flask with the tongs), attention may now be returned to melting the metal.

When the metal is at the correct casting temperature and the vacuum gauge registering a reading in the region of 27" Hg, the molten metal should be poured into the flask. It is useful to let the flame from the torch play on the residual sprue button for a few moments before releasing the vacuum.

A period of not less than two minutes should be allowed for the metal to cool and solidify before the flask is removed from the casting platform.

Casting with the water pump unit

It will be seen from Fig. 99 that the vacuum chamber has a gauge and two valves, or stop cocks, fitted to the casing. One valve will control the air supply from the vacuum chamber and, for convenience, let us call it A, whilst the other, B, will control the air supply to the casting head.

With valve A open and B closed, the water pump is activated simply by turning on the mains cold water supply. When the gauge shows a reading of 20–25″ Hg (manufacturer's figures), the flask is taken from the furnace and placed on the casting head. The metal is then melted in the crucible depression, the sprueing technique being the same as for pressure casting.

With the casting metal fully molten and spinning in the crucible depression, valve B is opened, thus releasing the vacuum, with the result that the molten metal is 'sucked' into the mould.

It is worth remembering that a minimum mains water pressure of 20 lbs/in² is recommended for efficient functioning of a water pump. Eventually it may produce sufficient vacuum on less pressure but will naturally take longer. It is prudent, therefore, to check your mains water pressure before purchasing such a machine.

Vacuum assist casting will produce sound, dense castings which are no better and no worse than those achieved by centrifuge. The main advantage claimed by the manufacturers of the motorised machine would seem to be the ability to cast flasks of greater capacity.

CASTING OF LOW TEMPERATURE ALLOYS

Up to this point description, in the main, has been geared to the casting of silver and gold alloys. There remains, however, one important area not yet covered, viz: the casting of low temperature alloys used widely in the manufacture of costume jewellery.

The alloys are usually of the lead/tin variety, having melting temperatures between 180° and 250°C. After casting, the pieces may be plated with silver, gold, or rhodium.

Whilst in no way could it be termed investment casting, there are many similarities in mould making and casting techniques that make the process worthy of inclusion.

For obvious economic reasons it would not be

feasible to cool an invested mould to a low enough temperature, when lead/tin alloys could be cast without fear of the metal being adversely affected by the higher temperature investment. On the other hand, manufacture of metal moulds would be prohibitively high, if not impossible in some cases. For this reason, the metal is poured into pattern spaces through a sprue hole in the centre of a revolving two-piece thick rubber mould.

In the first stage of the process the patterns, usually made of brass, are placed along with aligning pegs, between two discs of synthetic rubber. The whole assembly is then fitted between the heating platens of a hydraulic vulcanising press, similar to the model illustrated in Fig. 129. A pressure of between 10 and 15 tons per square inch is applied, the vulcanising heater switched on, and the moulds allowed to 'cook' for approximately one hour at a temperature in the region of 160°C. The temperature and time of vulcanising will naturally depend upon the thickness of the mould discs and the type of rubber used.

After vulcanising, the moulds are taken from the press, separated and the patterns removed. The flexible nature of the rubber obviates the necessity for any pattern draft allowance. Then, using a sharp, moistened knife, gates are cut from the sprue-former hole to the points of the mould cavities. The gates should be cut progressively wider and deeper as they near the sprue.

The moulds are then aligned and placed on the turntable within the centrifuge casting frame, where clamps are provided to align and hold the moulds during rotation. The protective flap is closed down and the machine turntable switched on.

With the turntable revolving at a predetermined speed, between 450 and 1200 r.p.m., depending upon the diameter of the disc mould and thickness of casting required, a measured amount of metal is ladled from the thermostatically controlled heating pot (usually situated to hand beside the machine) and poured into the funnel shaped depression clearly seen in Fig. 130. Centrifugal action, caused by the rapid rotation, will force the metal into the finest detail of the mould cavity.

The machine is then stopped, the mould removed and the casting extracted, the whole process taking approximately one minute.

When fettled, the castings may be given a light

Fig. 129. A heavy duty electrical press for vulcanising circular rubber moulds. Courtesy of N. Saunders Metal Products Ltd.

Fig. 130. A centrifuge casting machine for low temperature alloys. Courtesy of N. Saunders Metal Products Ltd.

102

Fig. 131. A typical lead/tin alloy centrifugal casting. Courtesy of G. W. Lunt and Son Ltd.

burnishing by barrelling with steel balls, and finished by bright plating.

A typical lead/tin casting illustrated in Fig. 131 formed part of the Times Newspapers Ltd promotion for the Tutankhamun exhibition at the British Museum.

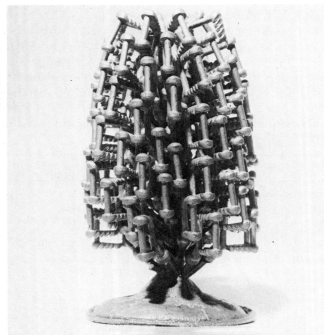

Fig. 132. An impressive tree casting of bracelet links by a high frequency casting machine.

Fig. 133. Sterling silver ring casting. Note the effective method of sprueing to get the maximum number of castings on the sprue button.

Casting defects

The amateur craftsman will be fortunate if he can consistently produce fault-free castings from the word go. Probably a high success rate will be achieved during his first dozen or so casts of an elementary nature, when he will take every precaution and work through the complete process to the letter.

Using correctly formulated materials and with equipment functioning satisfactorily, the mishaps may occur when, flushed with early success, he takes short cuts somewhere during the process. Sometimes these short cuts may be justified but, in the long term, the success rate is bound to be consistent with the degree of attention paid to correct procedures, and the ability to diagnose the cause of casting defects.

The following table may help the beginner with his casting problems:

Casting defects	Possible causes
Shrinkage in the heavier sections	1. Pattern incorrectly sprued. 2. Mould temperature too high. 3. Metal temperature too high at time of casting.
Incomplete casting	1. Incorrect positioning of pattern. 2. Pattern incorrectly sprued. 3. Burnout not completed. 4. Mould temperature not high enough. 5. Metal not hot enough when cast. 6. Insufficient metal. 7. Flask quenched too soon after casting.
Inclusions in the casting	1. Sharp corners at sprue extremities. 2. Flask not de-scaled from a previous firing. 3. Insufficient investment setting time allowed. 4. Initial heating of the flask too high, causing wax to boil. 5. Careless handling of the flask causing foreign matter to enter the sprueing system prior to casting. 6. Dirty crucible or casting metal.
Nodules or growths on casting	1. Incorrect positioning of the pattern causing air to be trapped under hollows. 2. Pattern not coated with wetting agent. 3. Wetting agent not allowed to dry before investing. 4. Air bubbles not eliminated from the investment slurry. 5. Invested flask subjected to vacuum for too long a period.

104

Casting defects	Possible causes
Rough surfaces on the casting	1. Poor surface finish on the pattern. 2. Wetting agent not allowed to dry before investing. 3. Incorrectly proportioned investment mix. 4. Insufficient setting-time allowed. 5. Initial furnace temperature too high, causing wax to boil. 6. Too high a percentage of 'once-cast' metal in the melt.
Fins or flash on casting	1. Pattern incorrectly positioned. 2. Incorrect investment/liquid ratio. 3. Investing time over extended. 4. Flask disturbed during initial setting. 5. Careless handling of the flask. 6. Invested flask heated too rapidly. 7. Mould temperature too low during casting.
Discoloured castings	1. Mould temperature too high when casting takes place. 2. Metal overheated. 3. Incorrect type of flame when melting metal. 4. Insufficient or incorrect type of flux.
Porosity or fine 'pin-holes' in the casting (Fig. 134)	1. Pattern incorrectly sprued. 2. Burnout not completed. 3. Metal overheated prior to casting. 4. Mould temperature too high. 5. Incorrect type of flux. The use of borax powder sometimes causes this condition. 6. Too high a percentage of 'once-cast' metal in the melt. 7. Too great a depth of investment above the pattern preventing the gases from venting off easily. 8. Crucible not preheated.

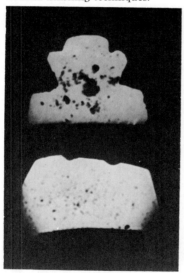

Fig. 134. Two examples of internal porosity in cast rings. The upper section shows massive cavities, possibly causing structural weakness. The lower example is, more or less, acceptable, provided that the small cavities are confined to within the body of the casting and not on or near the surface, where they could be exposed by normal finishing techniques.

Cleaning the casting

When casting has been completed it is usual to allow the flasks to cool for a short period of time in order to minimise crystallisation of the cast metal. The following times are suggested as a guide:

White gold and platinum—about 10 minutes.
Yellow gold and silver—until the sprue button loses its redness under normal workshop lighting.

The hot flask should now be plunged into a stout container of cold water to disintegrate the investment and reveal the casting. It is worth noting here that it is not advisable to carry out this operation in the sink, as the resulting slurry will rapidly clog the waste trap and may eventually block the drainage system.

Where casting is done in any quantity it would be wise to plumb-in a series of sludge traps and settling tanks. The process may be speeded up if a small amount of one of the proprietary agents, usually resembling a clear viscous liquid, is mixed with the milky waste.

The particles of investment held in suspension will rapidly flocculate and sink to the bottom of the container in a fraction of the time it would take under normal conditions. Disposal of the residual sludge is made easier if it is mixed with plaster of Paris and cast into blocks. Alternatively, the residue may be dried and put into polythene sacks for disposal.

It is inevitable that small amounts of investment—sometimes glazed—will remain in hollows and fine details of the casting. Commercial casters generally soak their castings in a strong solution of hydrofluoric acid to dissolve remaining investment and remove oxides from the surface of the metal.

Whilst this may be a standard trade practice where time is at a premium, the use of this acid is *not* recommended for the amateur. Its highly corrosive nature presents storage problems, whilst the fumes given off are highly toxic. Burns from acid splash can be very painful and reluctant to heal if not treated immediately.

With time on his side, it is far wiser for the amateur to persist with scrapers, probes and vigorous brushing to remove the surplus investment, and although every care may have been taken with burnout and casting, the odds are that there will be at least some surface oxide visible on the casting and sprue button, and pickling in a less toxic acid solution is indicated. See Appendix IV for details of appropriate acid solutions.

Pickling may be carried out in a suitably sized acid-proof container. An ovenware 'Pyrex' basin or dish is ideal for the purpose, though electrically heated pickling pots are obtainable (Fig. 135).

The solution may be used cold, though more rapid oxide removal is achieved if it is heated to about

Fig. 135. A small electrically-heated acid pickle bath. Courtesy of Hoben Davis Ltd.

80°C, ensuring that a suitable iron wire gauze with an asbestos centre is placed between the heat source and the container. Sudden thermal shock could cause the dish to shatter and spill its corrosive contents.

The casting, still joined to its sprue button, is carefully placed in the heated acid solution using a small pair of tongs or tweezers. These should be made of plastic, or have quartz tips to avoid contaminating the solution. Photographic print forceps are an ideal low cost alternative.

As soon as the metal loses its discoloration it should be transferred to a dish containing a suitable acid neutralising solution. Simply rinsing the casting under a running tap is not enough. Every casting has some degree of porosity, even though it may not be visible to the naked eye, and failure to neutralise the acid in these fine cavities may lead to corrosion and discoloration problems later on.

In order to keep acid pickles and neutralising solutions clean, when not in use they should be returned to clearly labelled bottles fitted with secure stoppers, and stored in a safe place.

When the pickle or neutralising solution becomes discoloured it should be discarded and a fresh mix made. A certain amount of care is required in the preparation of a fresh pickle if burns from acid splash are to be avoided. Mixing is best done in a sink, where any accidental spillage may be quickly flushed away.

Preparation of an acid pickle
A measured amount of warm water is placed in the dish and the concentrated acid is slowly poured down the inside wall of the dish. Some bubbling or spitting may occur if the acid is poured too quickly. **On no account should the water be added to the acid.**

Fig. 136. Sequence for mixing a sulphuric acid pickle solution.

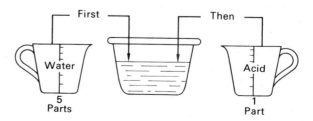

Fig. 136 graphically illustrates the mixing

107

sequence. In Fig. 137 both sequences are equally effective, though method 2 will prolong the life of the neutralising solution.

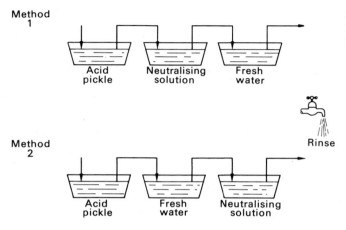

Method 1

Acid pickle   Neutralising solution   Fresh water

Method 2

Acid pickle   Fresh water   Neutralising solution

Rinse

Fig. 137. Alternative methods of removing surface oxides with pickling solutions.

When the casting is dry from the final rinse it should be separated from the residual sprue button. Side- or end-cutting nippers may be used and, though favoured by commercial casters, they do tend to leave a portion of the sprue still attached to the casting.

Although much slower, a piercing saw or jeweller's parting saw will allow the user to cut much nearer the casting and, as a consequence, minimise the amount of metal lost in final shaping—a point the amateur should not overlook when working in precious metals.

Suggested materials and equipment

METAL PREPARATION
Casting metal
Graphite or plumbago crucible for metal refining, one for each type of metal
Flux
Charcoal blocks
Acid pickles, hydrochloric and sulphuric
Acid neutralising solution
Non-ferrous pickling tongs
Accurate scales or balance for weighing metal
Metal melting torch: gas/compressed air, propane, oxy-coal gas, or oxyacetylene

108

CASTING—PRESSURE
Solbrig or air pressure caster
Metal melting torch
Flux
Heat resistant work surface
Tinted spectacles or face shield
Flask tongs
Asbestos gloves
Container of cold water for quenching flask

CASTING—CENTRIFUGE
Vertical or horizontal casting machine—the latter
    should be fitted inside a safety splash guard
Ceramic crucibles to suit casting machine, one for
    each type of metal
Asbestos paper for lining crucible (optional)
Flux
Metal melting torch
Flask tongs
Asbestos gloves
Tinted spectacles or face shield
Container of cold water for quenching hot flask

CASTING—VACUUM
Vacuum casting machine
Mains water supply of not less than 20 lbs/in² for
    water pump model
Plumbago or graphite crucible for heating and pour-
    ing the metal
Metal melting torch
Flux
Asbestos gloves
Tinted spectacles or face shield
Flask tongs
Container of cold water for quenching hot flask

CASTING—HIGH FREQUENCY INDUCTION
Casting machine
Availability of mains water supply for cooling
    purposes
Availability of mains electrical supply, 440 V, 3 phase,
    50 Hz, or 230 V, single phase, 50 Hz
Melting crucible to suit metal
Flask tongs
Asbestos gloves
Container of cold water for quenching hot flask

CLEANING THE CASTING

Basic
Assorted metal probes
Stiff nylon scrubbing brush
Scouring powder
Acid pickling solutions
Acid and heat resistant dish for pickling
Plastic tweezers
Neutralising solution
Bunsen burner
Tripod stand
Wire gauze with asbestos centre

Adjunct
Sand-blasting machine

9 Gemstones and findings

Gemstones

It is not within the scope of this work to describe the cutting and polishing of gemstones as it has been admirably covered in *Discovering Lapidary Work.*★ On the other hand it is important to note that never before has the craftsman been so fortunate in the wide range of gemstones currently available from lapidary dealers and gemstone merchants. There is literally a fantastic choice of facetted and cabochon cut stones, both natural and synthetic, in calibrated sizes and endless variations in tumble polished stones.

Where expense alone may rule out the possibility of obtaining a large, high quality, facetted precious gemstone, the amateur may now purchase, quite cheaply, a synthetic stone which is virtually indistinguishable from its genuine counterpart.

Whereas, formerly, simulated stones were made from glass or other substances and were relatively soft, the synthetic stone has the same chemical composition, hardness, light fraction, brilliancy, specific weight and dispersion factor as the genuine stone.

A genuine stone is the result of natural formation and influences, whilst the synthetic stone is man-made almost exclusively by the Verneuil process. Pure oxides, in the pulverised form, are subjected to a temperature in the region of 2100°C, causing crystallisation. The crystals will reach a size of some 200 carats (40 grams) in 6–8 hours.

Once cut and polished, it is extremely difficult, if not impossible to differentiate between a synthetic and a genuine stone, though expert analysis will not leave any doubt.

★ By John Wainwright, Mills and Boon, 1971.

In spite of the vast range of colours available, synthetics fall into two main classifications, the **spinels** and the **corundums**.

The synthetic **spinel** is crystallised magnesium—aluminium oxide of hardness 8 on the Mohs scale. The most popular spinel colours resembling their genuine counterparts are white, blue, aquamarine, tourmaline, chrysolite, and blue zircon.

Technical specification

Chemical formula	$MgO_3Al_2O_3$
Crystallization	Cubic
Hardness	8 (Mohs)
Specific weight	ca 3.60
Light refraction	ca 1.73

The synthetic **corundum** is crystallised aluminium oxide and may resemble ruby (red and pink), sapphire (golden and blue), alexandrite, garnet and amethyst.

Technical specification

Chemical formula	Al_2O_3
Crystallisation	Rhombohedral
Hardness	9 (Mohs)
Specific weight	ca 3.95 to 4.0
Light refraction	ca 1.76

Fig. 138. A combination lapidary machine. Courtesy of Highland Park and Ammonite Ltd.

Fig. 139. A compact combination lapidary machine. Courtesy of P.M.R. Ltd.

Plate III illustrates some of the available colours and cuts in synthetic stones.

The craftsman does not have to rely upon the dealer to supply gemstones. In keeping with the current popularity of lapidary work, there is a wide range of machinery for cutting and polishing gem-

112

Fig. 140. Two lightweight
economically priced machines
from Wessex Impex Ltd.

Fig. 141. An unusual tumbler of
elegant design. Courtesy of
J. Owen Engineering.

Fig. 142. A small Australian
combination unit from Kernocraft.

stones, specially designed with the amateur in mind.
Figs 138–42 illustrate a small sample.

Where the worker has access to a lapidary
machine, not only will he be able to manufacture
calibrated stones of standard shape, he will have the
advantage of creating new exciting forms not available through stone dealers.

Tumble-polished stones should not be confined to
the 'stick on to purchased findings' kitchen industry
currently so popular with many housewives. The
baroque shape of these stones lends itself to cast
jewellery, their smooth contours blend well with the
free-flowing shapes so characteristic of wax.

Small stones may cost only a few pence each,
whilst the light duty 'do it yourself' tumble polishing machines may be obtained at most competitive
prices (Figs 143 and 144).

Whilst it is customary to set gemstones after the
piece has been finally polished, in certain circumstances they may be mounted in the original wax
pattern and cast in situ. Fig. 145 illustrates a typical
example where the stone is first coated with a
layer of investment prior to constructing the wax
cage in order to protect it from the shock of the
injected metal during casting.

There are, however, certain reservations with the
technique of gem inclusion and success cannot be
guaranteed, as flaws in the stone may cause it to
shatter or craze during the burnout stage.

Change of colour is another variable; many stones
of the agate family change colour when heated to
only a few hundred degrees Centigrade. It would be
wise, therefore, to set up a flask with small pieces
of gem material embedded in the investment and
subject it to a normal burnout cycle.

Provided the temperature rise is slow and controlled, once the flask temperature has reached
700°C, the furnace may be switched off and the
flask allowed to cool down gradually within the
muffle. Under no circumstances should one attempt,
through all consuming curiosity, to accelerate the
rate of cooling, either by leaving the furnace door
open or by quenching the flask in water. The sudden
change in temperature would almost certainly cause
the stones to shatter.

Setting of gemstones
There are literally dozens of ways of securing a stone

113

to a piece of jewellery. The two most popular types are the box setting and the claw setting. Box settings are commonly used with cabochon cut stones which do not have light entering the base of the stone. The walls of the box are made from a thin section strip of metal soldered to form a loop the size and shape of the circumference of the stone. To achieve an exact fit the loop should be made a fraction of an inch smaller than required and should then be stretched to size on a stake with a few taps of a planishing hammer (Figs 146 and 147). This box is then soldered to a base of some sort. It may be a piece of flat sheet metal or it may be the cast form itself (Figs 148 and 149). The stone is secured by pushing the setting on to it by means of a piece of steel rod held in a handle of some sort. The end of the rod should be roughened slightly to stop it slipping, and the pressing over of the walls of the box should proceed alternately from opposite sides (Fig. 150).

When setting a square stone the corners should be set over first to prevent a wrinkle of metal accumulating at a point where it cannot be removed. The pressure is applied by hand at all points around the stone until a tight fit is achieved. The setting may be finally bedded down on to the stone with a few light taps from a chasing tool and hammer, or by means of a polished steel burnisher. If the stone is transparent or translucent, the inside base of the box should be polished, or lined with foil, to help reflect the light back through the stone.

Fig. 143. A low cost miniature tumbler. Courtesy of Ammonite Ltd.

Fig. 144. A mini tumbler from Kernocraft Rocks and Gems Ltd.

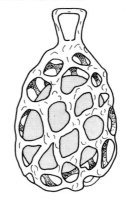

Fig. 145. A 'cage' pendant with a gemstone cast in situ.

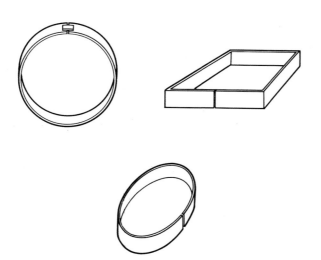

Fig. 146. Simple box settings.

114

Fig. 147. Stretching the box to
size on a square triblet.

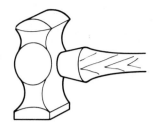

Fig. 148. The box is soldered to a
base of some sort.

Fig. 149. A circular box setting
soldered to a cast ring.

Fig. 150. A rub over or burnished
setting.

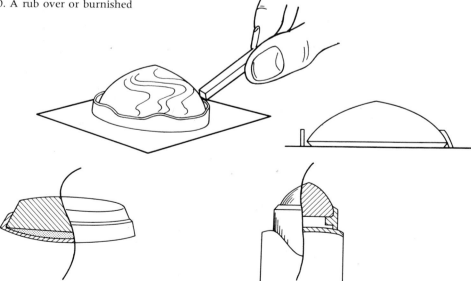

Claw settings the shape of a miniature coronet can be purchased from a bullion dealer, ready made with a bezel, for the standard sizes and cuts of stones up to about 8 mm diameter (Fig. 151).

Where the stone is of an unusual size or shape, this kind of setting can be easily made in the following fashion:

Form a tube somewhat narrower than the girdle of the stone by soldering a seam in a strip of thin metal. By means of a collet plate and punch form it into a cone just large enough to support the stone (Fig. 152). With a scorper or setting burr, cut a rebate in the mouth of the cone to support the stone (Fig. 153). Next, file eight 'V' shaped grooves around the outside circumference of the collet to leave the claws standing up. File a further eight grooves around the base to interlock with the previous set. The circular bezel upon which a setting of this kind normally stands is made from square section wire soldered into a circle. The tips of the claws are burnished over to secure the stone. The same basic techniques can be used to form oval, rectangular, or other shaped settings.

Fig. 151. A crown or coronet setting with matching bezel.

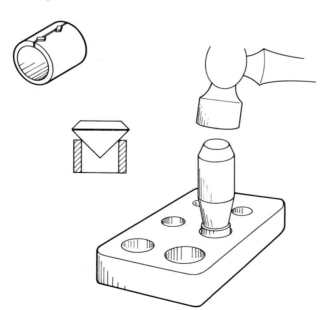

Fig. 152. Forming a coronet setting.

Another form of setting commonly used on casting is the beaded setting (Fig. 154). A recess is cast or drilled into the body of the piece, the shape of the stone and deep enough for the girdle of the stone

116

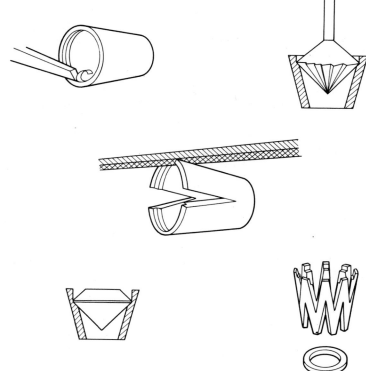

Fig. 153. Final shaping of the setting.

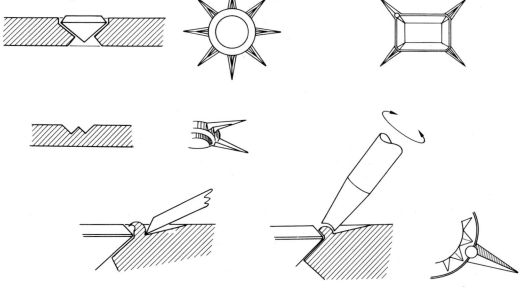

Fig. 154. Stages in making a beaded setting.

to sit just below the surface. Arranged symmetrically around the stone, if it is circular, or at the corners if it is octagonal, a series of 'V' section grooves are cut with a scorper. In the base of each groove a small inverted 'V' section is left upstanding. A sharp scorper is used to curl the ridge of metal over the lip of the stone. A beading tool is used to burnish this chip of metal down on to the stone. There are no second chances with this type of setting, so practice is advised before it is used on a valuable piece of jewellery. As a technique it is best suited to use with harder metals like platinum; used on silver, larger beads will be necessary.

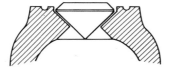

Fig. 155. Simple paved settings.

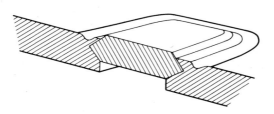

Another form of setting commonly used on cast forms, particularly signet rings, is shown in Fig. 155. The grooves surrounding the stone may be cast or subsequently cut with a graver.

The settings so far described are designed, in the main, for securing stones that have been cut to a regular standard form. They can be adapted to deal with stones of somewhat different shapes but where tumbled stones, pebbles, slabs, crystals and mineral

118

specimens are involved, special setting techniques are called for.

Lost wax casting techniques are particularly suited to setting this type of stone since claws can be formed in wax to the exact profile of the stone. The whole model, including the stone, is immersed in warm water to soften the wax and the stone is gently removed from the claws. The pattern is cast with the claws in the open position (Fig. 156).

Fig. 156. Carving a claw setting in wax with the stone in situ.

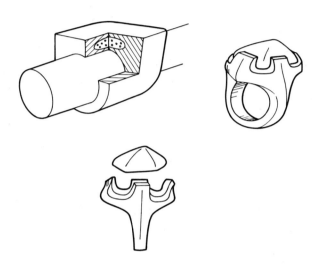

Fig. 157. Peg fitting of irregular shaped stones.

Irregular shaped tumbled stones and crystals can be secured to roughened metal surfaces with epoxy resin adhesives, and where additional security is required the stone may be drilled with a diamond tipped drill and secured on a peg (Fig. 157). The classic setting for this kind of gem was developed for the setting of pearls. A round peg of metal is split

119

down the middle and soldered to a supporting cup, or directly to the body of the piece. A tiny wedge is just inserted into the bifurcated peg. The peg and wedge are inserted into a hole drilled into the stone or pearl and containing a little adhesive. The hole is broader at the inside end than at the mouth and just deep enough to take the peg when the wedge is driven home (Fig. 158).

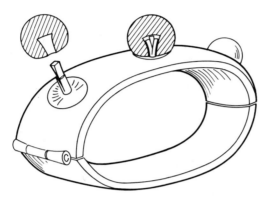

Fig. 158. An alternative method of peg fitting.

Findings

Findings are all those fastening and joining devices used in jewellery making. These items are usually attached to the main body of the piece once the casting is complete and just before the finishing touches and final polish.

If the item is made of silver or gold and is to be assayed then the findings must be of the same metal and must be attached by silver or gold solder of the same assaying quality.

Silver and gold soldering is a vital element in the jeweller's craft and though the process is basically simple it does require a certain care, precision and dexterity. Gold and silver solders can be bought from the bullion dealers in five different grades*— 'enamelling', 'hard', 'medium', 'easy' and 'extra easy'. The names indicate the melting points of the solders and when making a series of joints in close proximity one works down the scale of hardness so that the heat required for the later joints does not cause the earlier ones to run. With careful control of the flame, and by masking earlier joints with a paste of

* See Appendix for details.

120

powdered rouge and water, many more than five joints can be achieved. For effective results the surfaces to be joined must be bright and free from grease, scale, or other dirt. They must be touching, or very nearly so, and should be coated with a milky flux, either of borax and water or one of the proprietary fluxes. The solder, which also should be clean and bright, may be applied to the joint in a number of different ways.

One of the simplest and most effective methods is to cut a fringe on the end of the sheet of solder, and then with one or two parallel cuts at right angles, chop off a number of square snippets, or paillons (Fig. 159). As many of these paillons as are needed are then placed in position on the fluxed joint with a pair of fine tweezers, or the tip of a brush (Fig. 160). The heat from the gas blow-torch is applied gently at first until all the water has evaporated from the flux. Once the flux has passed through a white and frothy state the heat may be intensified until the metal flows into the joint.

Where small delicate findings are to be soldered to relatively massive pieces of jewellery, it is vital that the larger piece is heated thoroughly before the flame is allowed to contact the smaller one. If this is not done the tiny finding will heat up too fast, the solder will fuse to it, but not to the larger piece. Worse than that, it may melt away before the temperature of the other piece has risen sufficiently for the joint to be made.

Fig. 159. Cutting paillons of solder.

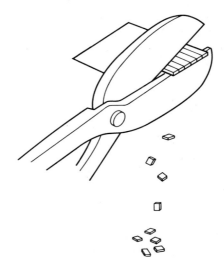

121

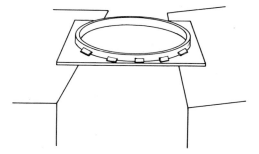

Fig. 160. Position of paillons prior to heating.

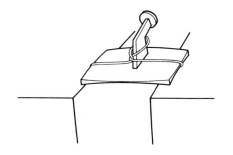

Fig. 161. Soldering a small finding.

The soldering may be done on a hearth of asbestos or fire brick, with the job surrounded by pieces of fire brick in such a way as to reflect the heat back on to it. To avoid direct contact between the small finding and the flame, it may be necessary to heat the piece from the other side. This is easily achieved if the job is supported on a wire mesh or between two fire bricks (Fig. 161).

When the joint is complete, the excess borax is dissolved away in the acid pickle and any surplus solder removed with files and emery. It is a common mistake with beginners to use too much solder and so produce a messy joint. When the quantity needed is well judged, capillary attraction of the joint will take up all the solder, leaving no excess.

When very small quantities of solder are required, it may be applied in the form of filings in a medium of flux. When large amounts of solder are needed to fill the joint, it may be applied in a long strip, held by corn tongs and run into the joint once it has been raised to the required temperature. This takes some skill to manage effectively, but allows greater control over the amount and placing of the solder. Failure in soldering is most likely to be the result of too wide a joint, dirty, oxidised surfaces and/or uneven heating of the two halves of the joint.

Brooch fastenings

A number of different solutions have been found to the problem of securing a brooch to fabric. Commercially made joints and catches are simple and effective. They should be attached to the brooch so as to give as long a pin as possible, with about two-thirds of the weight hanging beneath the pin, which should be worn horizontally. The catch should be attached so that the opening faces towards the bottom of the brooch, and the joint aligned so that the pin points a few degrees above the catch. Soft iron binding wire threaded through the holes and bent to form a handle will assist the placing of the joint and catch during the soldering process (Fig. 162).

Fig. 162. A commercial brooch finding.

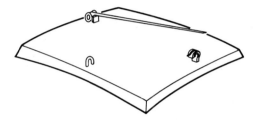

When working in silver, the pin should be made from nickel-silver wire and should not be attached until after the piece has been assayed. The loop on the end of the pin can be made by winding the wire round a small nail held in the vice. The excess wire is sawn away and the loop aligned with a pair of taper nosed pliers. A tiny snippet of solder will secure the joint and a tap with a hammer will flatten and harden the loop. The shank of the pin may be made hard and springy by giving it a twist through 180° with a pair of pliers. The pin is secured to the joint by means of a rivet of silver wire tapped into the countersunk recesses at either end of the hole by means of two hollow faced punches (Fig. 163). The pin should strike the positive stop of the joint before reaching the catch and should, therefore, be under slight tension when worn.

A simpler and cruder pin and catch may be fashioned from wire alone. A loop of three-quarters of a turn, or 270°, is put into the pin, which is soldered directly to the brooch with the point at 180° to the catch. The soldering removes the springiness from the pin but it is replaced as the pin is bent over

123

into position near the catch (Fig. 164). The catch is no more than a piece of wire in the shape of a question mark.

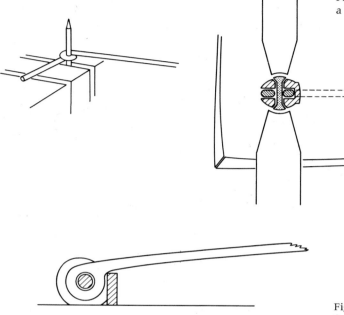

Fig. 163. Forming and securing a brooch pin.

Fig. 164. Alternative method of forming the pin.

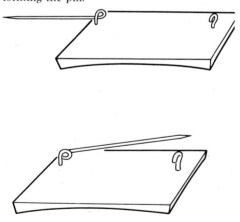

A much more substantial joint may be made from tubing and a piece of sheet metal folded into an angle (Fig. 165). A 'V' block, or jointing tool, should be used to file the ends of the tubing at right angles. The vertical face of the angle supports the tube and acts as the positive stop for the pin. The brooch pin is soldered to a slight groove at a tangent to the middle of the centre piece of tube. The two outside pieces are secured with the minimum of solder necessary and the centre piece is filed down to a close fit. The hinge pin is inserted in the way described above. When fixing the joint and catch to a brooch, it is no trouble to solder a small semi-circular hoop in position as an attachment for a safety chain (Fig. 166).

Chain making

Safety chains, and chains for cuff-links, necklets and bracelets can be bought from various dealers ready made, but when necessary they can be made with little difficulty from wire. File a steel rod to whatever section is required for the shape of the individual links. Anneal the wire and wrap it in a tight spiral

124

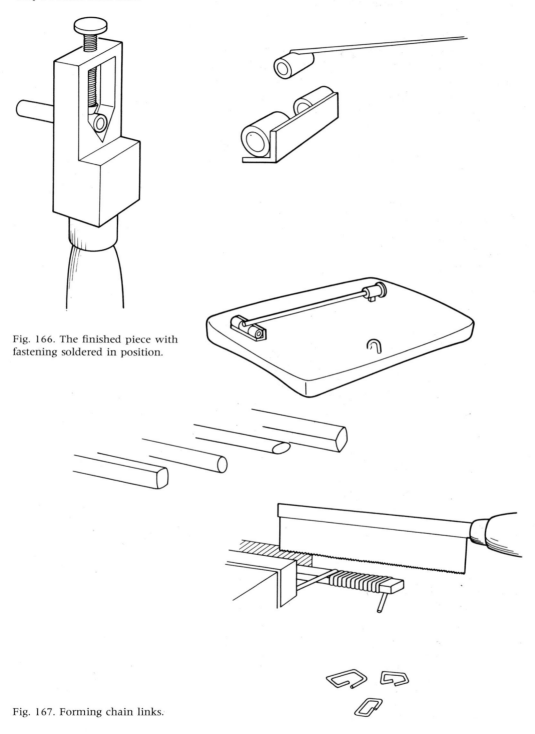

Fig. 165. A stronger type of fitting may be made from tube.

Fig. 166. The finished piece with fastening soldered in position.

Fig. 167. Forming chain links.

round the former. With the steel former held in a vice, use a parting saw to cut along the length of the spiral and slide off the links as they are cut (Fig. 167). Each link will need to be cleaned up, hooked on to the chain and given a slight twist to bring the ends of the wire into line. To solder the links, the chain must be held securely so that the joints are not in contact with other links, or else the whole thing will fuse solid. One method is to stake the chain out between two pins stuck into a soft asbestos pad. Each joint is painted with flux and a minute piece of solder is put in position with a brush or tweezers. A fine pointed flame is essential for chain work.

An alternative method of holding each link while it is soldered separately is to secure it with two pins to a piece of asbestos held vertically between two fire bricks. A pin for a safety chain may easily be formed from a single piece of wire, using a pair of fine taper-nosed pliers (Fig. 168).

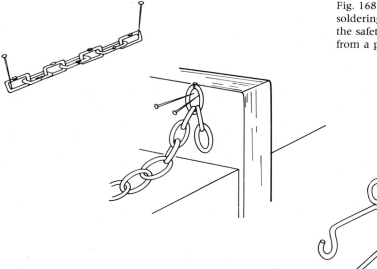

Fig. 168. Two methods of soldering the links. The pin for the safety chain may be formed from a piece of wire.

Fastenings and catches

There are literally hundreds of devices to secure the ends of belts, bracelets, and necklets. Some of them are sold as standard fittings by dealers, others must be made by the craftsman himself. A selection of devices is shown in Fig. 169.

Earrings

The simplest kind of earring is that designed for

126

Fig. 169. A selection of methods for joining the ends of a chain.

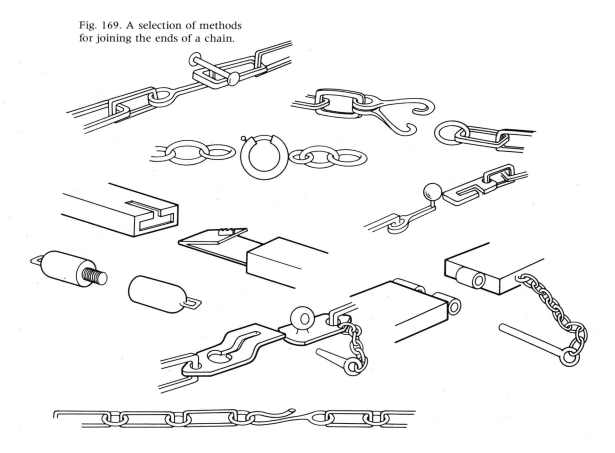

Fig. 170. Ear wires and butterfly clips.

wearing in pierced ears. Ear wires and butterfly clips can be made but are probably better purchased (Fig. 170).

Screw clips can also be bought. They should be soldered to the body of the earring while the wire is straight and then bent up into position to harden it once more. A small dome-shaped cap, soldered in position over the end of the wire will make them more comfortable to wear (Fig. 171).

Ear clips can be either bought or made. As they are very dependent on the hard springiness in the moving part, it is important to remove it while the base is soldered to the body of the earring (Fig. 172).

Cuff-links

A number of different methods have been employed to join the two halves of cuff-links. The traditional short chain is still the most comfortable to wear,

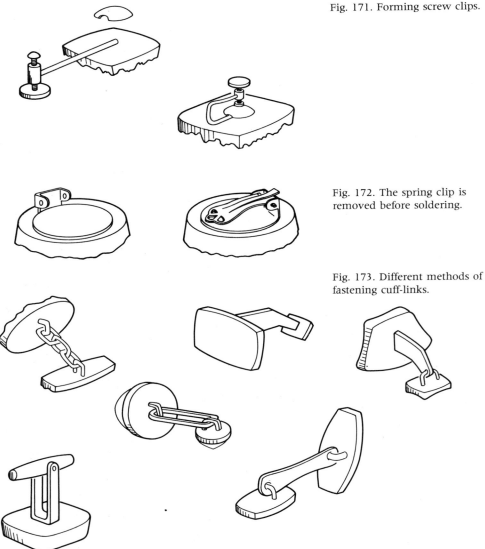

Fig. 171. Forming screw clips.

Fig. 172. The spring clip is removed before soldering.

Fig. 173. Different methods of fastening cuff-links.

though the more rigid devices are quicker and more convenient to thread through the button hole (Fig. 173). The total distance between the two halves of the link needs to be about two centimetres, and when soldering the joining piece in position remember that the button hole is cut parallel to the end of the sleeve and the link should be aligned accordingly.

Rivets, cheniers and hinges

Riveting may sound a somewhat crude method of fastening for jewellery work, but there are many

Fig. 174. Rivets as a decorative
feature.

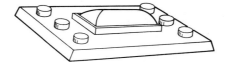
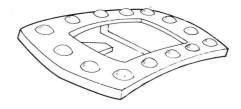

instances when using materials that will not withstand heat, such as bone, leather, plastic or wood,
in which it is the best means available. Domed or
cheese head rivets can be used as a feature of the
design, lending an illusion of strength and massive-
ness to the piece (Fig. 174).

A rivet is nothing more than a pin of metal holding two or more sheets of material together. The
head of the rivet may stand proud of the surface, or
it may be countersunk into it. When making rivets
from wire, file a very slight taper on the wire and
run it into the hole as far as it will go. Cut off the
excess length, leaving sufficient metal standing proud
of the hole to form the rivet head. With the wider
end below, the length of wire standing above the
surface is rounded with light blows with the ball of
a hammer.

The job is supported either on a flat metal surface
or on the concave face of a setting punch. Where the
rivet head must lie flush with the surface the hole
should be countersunk and the rivet head be
hammered down into the hollow. Any surplus
metal is filed away until a flat finish is obtained
(Fig. 175). If one, or both, of the sheets to be joined
is a soft material, such as wood or leather, a metal
washer must be used to protect it from distortion.

Cheniers are essentially tubular rivets, but instead
of the head being formed by blows from a hammer
the end of the tube is spread out by means of a burnisher. This is of particular advantage where delicate
work is involved. For this reason, this method is
commonly used for attaching enamelled panels to
cups, shields and other large pieces. One end of the
tube is soldered with enamelling solder to the back

129

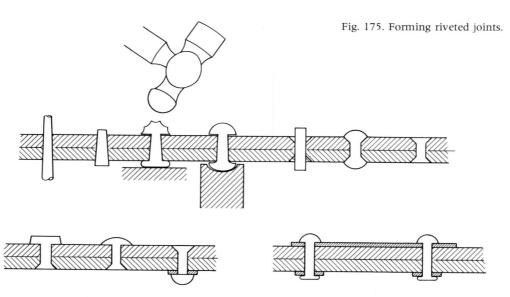

Fig. 175. Forming riveted joints.

of the panel before the enamel is applied. Holes are drilled in the body of the piece to receive these tubes and the finished panels are placed in position and the cheniers are burnished over (Fig. 176).

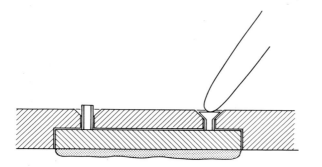

Fig. 176. An example of chenier riveting.

For hinge making, a jointing tool of miniature 'V' block is required to achieve absolute accuracy in filing the ends of the tubes to a right angle. Wire must be drawn down so that it fits snugly inside the tubing to act as the hinge pin. Where a silversmith's hinge is to be made, two tubes must be used that nest exactly one within the other. In the simplest type of hinge three, five, or seven short lengths of tube are laid end to end in a groove filed along the joint. These are tacked into position, alternately to either half of the joint, with a minute piece of solder on each. If too much solder is used, or if the flux is allowed to flow into the joint, the whole thing will fuse solid, so beware! Once the tubes are positioned

130

the two halves can be drawn apart and a complete soldering achieved.

The outside ends of the hinge tubes are counter-sunk, and the ends of the hinge pin are set over like rivets and filed flush. Alternatively, the ends of the hinge pin can be soldered to the end tubes, though this is a risky business (Fig. 177).

The silversmith's hinge, or box hinge, is similar except that the tubes are not attached directly to the two halves of the joint but sit in channels cut from a larger diameter tube. These channels are about 100° arcs of a circle, and provide a positive stop to the movement of the joint. For this reason they are used on good quality box work and also on holloware bracelets. It is this outside channel that is soldered to the two halves of the joint (Fig. 178).

Fig. 177. Tacking the tubes in a simple hinge fastening.

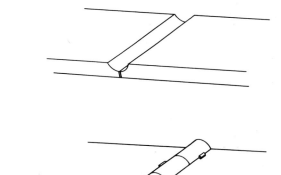

Fig. 178. The silversmith's box hinge.

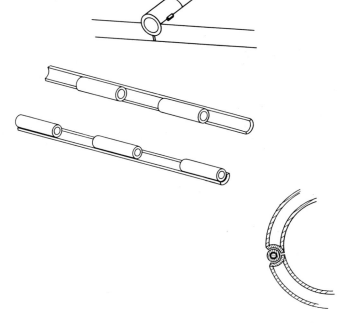

Suggested materials and equipment

Basic
General metalworking tools
Assorted riffler files
Assorted needle files
Piercing saw
Jeweller's parting saw
Jeweller's peg
Round, square, and oval triblets
Planishing hammer
Small rawhide mallet
Round nosed pliers
Taper nosed pliers
Ring pliers
Duck billed pliers
Assorted draw plates
Draw tongs
Doming block and punches
Setting block and punches
Mandrels for chain making
Assorted gravers
Jeweller's drill stock and set of small drills
Gem setting punches
Burnishing tools
Small soldering hearth
Lightweight soldering torch
Borax cone and dish or proprietary flux
Charcoal and asbestos soldering blocks
Acid pickles
Pickling trough or dish
Neutralising solution
Acid resistant pickling tweezers
A range of gold and silver assayable solders
Assorted jewellery findings in assayable quality alloys
Gemstones: Calibrated sizes of natural and synthetic
stones both faceted and cabochon cut
Assorted shapes and sizes of tumble-
polished and native cut stones
Clear epoxy resin cement

Adjunct
Lapidary machines and materials

10 Finishing techniques

Fig. 179. A compact polishing unit with flexible shaft and hand piece. Courtesy of Kernocraft Rocks and Gems Ltd.

Fig. 180. A combination pendant motor support and jewellers peg suitable for mounting in a vice.

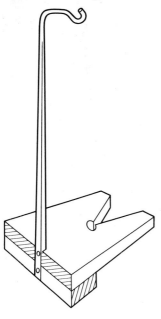

A well cast piece should require the minimum of hand working, other than removing the evidence of sprue junctions and the soldering of settings and findings, though the form and surface finish of the original pattern will obviously have some bearing on the amount of finishing work to be done.

Small nodules may be twisted away with a pair of taper-nosed pliers, or removed by careful use of fine cut files, scrapers, or gravers.

Imperfections in awkward-to-reach areas may be removed by a variety of abrasive points or wheels mounted in the handpiece of a flexible shaft. The shaft may be an extension of a small polishing lathe (Fig. 179) or can be powered by a pendant motor or suspension unit. The motor will need to be hung at a convenient height on a supporting arm; Fig. 180 illustrates such a device which, when gripped in the jaws of a vice, may be used in conjunction with a jeweller's peg.

Figs 181–3 show a small selection of flexible drives, hand pieces and suspension units currently available for amateur and professional use.

Sand blasting

When mounted points or abrasive wheels fail to clean up and brighten up awkward internal surfaces, the use of a sand blasting machine may be called for. Basically, this machine is a metal box requiring a compressed air supply which is directed to an internal nozzle.

Abrasive grits are picked up by the jet stream and directed on to the work which is held in a gloved hand. The glove, which is easily replaceable, forms an integral part of the machine to prevent the escape of grit. A glass window illuminated by a small electric light, fitted externally or internally, will allow the work to be seen during the operation (see Figs 184–6).

This type of machine does have one minor disadvantage in that unless the grade and type of abrasive material is selected with some care, several machines may be necessary if progressively finer finishes are required. This is particularly true where the machine may also be required for the removal of oxides and glazed investment from castings and sprue buttons, and to frequently change the grit for a different type or grade is not a practical proposition. However, a good range of abrasive materials is available, some in the form of corundum or quartz and silica grit which impart a bright matt finish, whilst others may be of the glass bead type which will burnish the work to a high surface lustre.

There may still be small scratches or surface imperfections on the work which have not been removed by the previous operations. It is important that these flaws are removed at this stage, as subsequent polishing will only serve to highlight them in contrast to surrounding areas. In certain cases gentle rubbing with a Water-of-Ayr stone, well lubricated with water, will prove to be effective.

Barrelling

Not widely known outside the trade, one commercial method of polishing jewellery is barrelling or burnishing. This technique, which is similar to the tumble-polishing of gemstones, is eminently suitable for amateur use and will achieve an excellent surface finish on a variety of work (Figs 62, 63 and 187).

The pieces to be polished are placed in a sealed, leak-proof container along with burnishing media, cleaning compound and water and, depending upon the diameter of the container, revolved at a speed between 35 and 50 r.p.m.

Although commercial machines are available for this process (Figs 188 and 189), and trade production requirements would obviously demand their use, one of the many small tumblers used in conjunction with a rubber, or rubber lined hexagonal steel barrel, will give comparable results at much lower cost.

The type of media most commonly used for burnishing is in the form of hardened steel shapes

Fig. 181. A pendant motor fitted with a flexible shaft, hand piece and variable speed foot control. Courtesy of Kernocraft Rocks and Gems Ltd.

Fig. 182. A suspension unit complete with flexible shaft, hand piece, support and variable speed foot control. Courtesy of John Quayle Mfg. Co. Ltd.

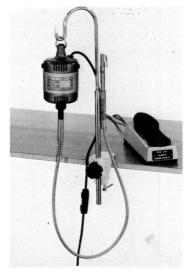

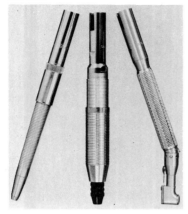

Fig. 183. Alternative hand pieces for use with the suspension unit. Courtesy of John Quayle Mfg. Co. Ltd.

Fig. 184. A small American sand blast unit with an illuminated lens window. Courtesy of J. F. Jelenko and Co. and Marcel A. Courtin.

(Fig. 190) which may be mixed together or, depending upon the form of the work to be polished, used separately. The shapes are sold by weight and are usually supplied with a coating of grease to prevent rusting.

It is essential that all traces of grease and dirt are removed before use, as work which has been barrelled in contaminated media often takes on a dull grey coating, completely lacking in surface lustre. The cleaning may be done in two stages:

1. A thorough cleaning with a suitable de-greasing compound, such as Jizer, followed by several rinses in hot water.
2. The shapes are now placed in the barrel to a level of between $\frac{1}{2}$ and $\frac{2}{3}$ of its capacity and only just covered with water. About a level dessert spoon of burnishing compound is added to the load. Canning's Gallay No. 8 is ideal for this purpose, though a small amount of liquid soap or detergent may be used as a slightly less effective alternative.

The lid should now be securely fitted and the barrel revolved for 10–15 minutes, after which time both the media and the inside of the barrel should again be thoroughly rinsed.

Both barrel and media are now ready for use and in order to prevent damage to delicate pieces, alternate layers of jewellery and media are loaded until the recommended level has been reached. For best results the amount of work should not exceed $\frac{1}{3}$ the volume of the media.

Water and burnishing compound are added in the same proportion as before and because of the considerable weight—a standard 127 mm diameter rubber barrel, fully loaded may weigh up to $2\frac{1}{2}$ kilos —a securely fitting lid is essential if subsequent leakage is to be avoided.

The work may be tumbled for anything between three and six hours, during which time the barrel may be removed occasionally so that the work may be inspected for quality of finish. It is prudent to stand the barrel in a washing-up bowl, placed in the sink, not only to avoid the mess caused by the suds, which may gush out when the seal is broken, but also to retain any of the shapes which may be accidentally spilled.

135

Fig. 185. A small sand blast unit. Courtesy of Metrodent Ltd.

Fig. 186. A bench model blast cleaner with an integral dust collecting unit. Courtesy of Guyson Industrial Equipment Ltd.

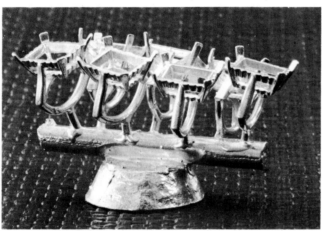

Fig. 187. A barrelled casting.

Fig. 188. A double ended barrel-polishing machine. Courtesy of Ferraris Engineering and Development Co. Ltd.

Fig. 190. Steel burnishing shapes (metallic media) used for barrel-finishing jewellery.

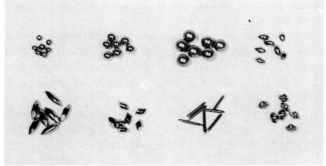

Fig. 189. A smaller commercial barrelling machine. Courtesy of W. Canning Ltd.

136

This technique has certain advantages over conventional lathe polishing. No metal is removed during the process so that surface detail is uniformly bright and crisp. Polishing mops and finger bobs, because of their shape cannot always get into the fine detail so some variation of surface finish must be expected.

The high safety factor is an added bonus with the elimination of risk to the work if it is snatched from the hand by a polishing mop—a not too uncommon occurrence with young pupils.

In between loads the metal shapes may be left for periods up to a week in water to which has been added a small quantity of burnishing compound or alkaline metal cleaner. When not required for longer periods, the shapes should be rinsed, thoroughly dried and stored in a sealed container along with a suitable de-moisturising agent.

Machine polishing

This process is carried out on a polishing lathe in conjunction with mops, brushing wheels, felt bobs and various polishing compounds. As strong pressures should not be applied—loss of surface detail will be inevitable if it is—a heavy duty machine is not necessary, though some lathes are fitted with a dust collection unit where a considerable amount of polishing has to be done. Figs 191 and 192 show two polishing lathes particularly suitable for jewellery work.

Depending upon the design of the machine, mops may be 'screwed' on to a threaded taper spindle nose or held between two washers or flanges and secured with a nut, the latter method offering the more secure fixing with minimum risk of the mop coming loose. For maximum safety in busy workshops, a taper spindle nose should be shrouded with a stationary metal sleeve when not carrying a mop.

A new mop should always be dressed when first fitted and is quite simply done by holding a clean wire brush or cylindrical cheese grater against the revolving mop. This not only levels the face but also causes the threads to become knitted together, giving a smooth velvet-like surface. The use of a mask or handkerchief tied around the face is advisable during this operation, as a considerable amount of fluff is released into the air.

The mop is charged by switching off the motor

and applying the polishing composition bar against the revolving surface in a gentle braking action. This method is more economical on the composition bar than charging at maximum revolutions.

Polishing is fraught with hazard if certain precautions are not observed. The 'underhand' method should be used so that the work is held at, or just below the horizontal centre line (see Fig. 193). The surface if the piece being polished is kept in line with the face of the mop and the movement should be at an angle to the mop as shown in Fig. 194.

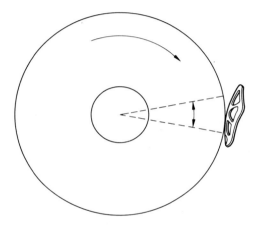

Fig. 193. The work should be held at or just below the horizontal centre line.

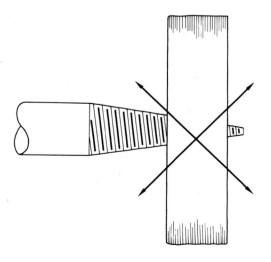

Fig. 194. The direction of travel should be at an angle to the mop surface.

138

Fig. 195. A 50 fold unbleached calico mop and Lustre composition bar suitable for grease mopping.

Fig. 196. A 40 fold finishing mop and Rouge composition bar suitable for colouring.

Fig. 197. A 24 fold 'Swansdown' mop and powdered rouge suitable for dry rouging.

The degree of pressure which may be applied will obviously depend on the form and construction of the piece being polished and, to a lesser extent, on the metal in which it has been cast. Whenever possible the work should be held on a piece of wood (see Fig. 49). Not only will this support the work, allowing even pressure to be applied, it will also avoid frictional heat being dissipated into the fingers.

General polishing may be simply classified under two main headings:

1. Grease-mopping which provides an acceptable surface polish,
2. Colouring which brings out the latent colour of the metal.

The piece is first polished on a mop, using say, a Lustre composition and when an acceptable surface has been achieved, any surplus composition, which may have been trapped in surface detail, is removed with a soft hand brush charged with a soap solution, and the model thoroughly dried.

The work is coloured with a rouge composition applied to a finishing mop. Dry rouging is sometimes preferred and with this technique the work may be polished on a 'swansdown' mop to which a little of the finest rouge powder, mixed into a paste with water or methylated spirit, has been applied with a brush or a piece of wood. A selection of suitable mops and polishing compositions is shown in Figs 195–7.

Bristle or soft nylon brushing wheels are often used for dry rouging intricate detail which might otherwise be blurred by a mop. For this and the previous method, the use of a simple wooden protective cowl (Fig. 200) will contain the main bulk of polishing composition thrown out by centrifugal action.

The same protective device may also be used when, as an alternative to sand blasting, a matt or satin finish is required. This finish calls for the use of very fine crimped steel or brass wire rotary brushes and wheels (see Fig. 201) which may shed a few of their bristles when revolving at speed.

Very little pressure should be applied to the wheel, the end of the bristles being allowed to strike the work quite lightly. On no account should a brass wire wheel be used on silver as it will quickly impart a yellow tinge to the surface of the work.

In some instances both lathe polishing and barrelling could damage very fine settings and the only suitable alternative method would be to use jeweller's polishing thread, charged with polishing composition. This method, though basically simple, requires a certain amount of care where crisp detail is to be retained.

Jewellery manufacturers may strip the surface layer of metal away in an electrolytic bath to remove any obstinate fire stains or flux deposits. Similarly they may deposit a fine layer of gold and silver to brighten the surface prior to polishing.

As both of these processes involve the use of potassium cyanide solutions which are potentially dangerous and pose problems of toxic waste-disposal in this pollution conscious age, their use is best confined to industry where rigid attention may be paid to safety precautions.

Fig. 198. A selection of some of the many felt wheels and polishing bobs available. Courtesy of W. Canning Ltd.

Fig. 199. A bristle wheel for polishing fine detail. Courtesy of W. Canning Ltd.

Fig. 200. A protective cowl will keep the polishing area clean.

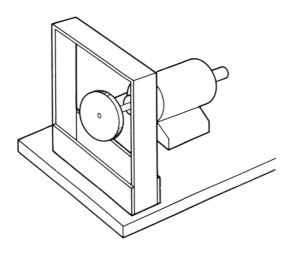

Fig. 201. A Vertex wire scratchbrushing wheel suitable for matt or satin finishes.

Suggested materials and equipment

Basic
Jeweller's parting saw
Piercing saw
Fine cut files of assorted section
Assorted needle files
Assorted gravers or scorpers
Straight and curved burnishers
Curved and three-square scrapers
Taper nosed pliers

140

Fine emery or silicon carbide paper
Water of Ayr stones
Registered stamps for Assay
Jeweller's polishing thread
Small polishing lathe
Assorted polishing mops
Mop dresser
Felt finger bobs
Polishing composition bars
Jeweller's rouge powder
Nylon or bristle brushing wheels
Fine crimped steel and brass wire wheels
Protective spectacles or goggles
Tapered wood dowel rods of assorted diameters
Wood strip for supporting work
Metal polish
Soft polishing cloths
Small gem tumble-polishing machine
Rubber or rubber-lined barrel
Metal burnishing shapes (Metallic Media)
Burnishing compound

Adjunct
Commercial barrelling machine with recommended
    media
Sandblasting machine
Suspension unit with support, flexible shaft and
    handpiece
Assorted mounted points, wheels and brushes

Appendix I.
Further reading

Practical Gemstone Craft, Helen Hulton, Studio Vista.
Cutting and Setting Stones, H. Scarfe, Batsford.
Discovering Lapidary Work, J. Wainwright, Mills & Boon.
Gemologist Compendium, R. Webster, N.A.G. Press.
The Amateur Lapidary, R. A. Jerrard, D. Bradford Barton.
Gem Cutting, J. Sinkankas, Van Nostrand.
The Art of Jewellery, Graham Hughes, Studio Vista.
Jewellery, Graham Hughes, Dutton Vista.
Modern Jewellery, Graham Hughes, Studio Vista.
New Design in Jewellery, D. J. Wilcox, Van Nostrand Reinhold.
Modern Jewellery Designs and Techniques, I. Brynner, Van Nostrand Reinhold.
The Amateur Jeweller, R. A. Jerrard, D. Bradford Barton.
The Goldsmith's & Silversmith's Handbook, Station Abbey, Technical Press.
Design & Creation of Jewellery, Von Neuman, Pitman.
Creative Gold and Silversmithing, Sharr Choate, Allen & Unwin.
Creative Casting, Sharr Choate, Allen & Unwin.
Four Centuries of European Jewellery, E. Bradford, Spring Books.
Lost Wax Air Pressure Casting, K. Edwards and G. Davies, Mills & Boon—Griffin Technical Series.
Centrifugal Casting by the Cire Perdue Process, C. Rosen. Contact Publications.
Lost Wax The New Modern Craft, Kerr Manufacturing Co., Obtainable from Hoben Davis Ltd.
A Handbook of Dental Laboratory Procedures, J. F. Jelenko & Co., Obtainable from Marcel Courtin.
Gems—The British Lapidary Magazine, By subscription only from 29 Ludgate Hill, London EC4M 7BQ.

Appendix II.
Buyer's guide

Dental sheet modelling wax (Toughened)	1. Amalgamated Dental Company (Claudius Ash & Co. Ltd.) 2. Ammonite Ltd. 3. Metrodent Ltd. 4. S.S. Dental White Ltd.
Model Cement (Sticky wax)	1. Amalgamated Dental Co. Ltd. 2. Ammonite Ltd. 3. S.S. Dental White Ltd.
Wax Profiles (including round section for sprues)	1. Ammonite Ltd. 2. Chaperlin & Jacobs Ltd. 3. Hoben Davis Ltd.
American Carving and Pattern Waxes	1. Hoben Davis Ltd.
Injection Waxes	1. Blayson Olefines Ltd. 2. Hoben Davis Ltd. 3. Alexander Duckham & Co. Ltd. (Trade only) 4. W. J. Hooker Ltd.
Water Soluble Wax	1. Blayson Olefines Ltd.
Inlay Wax	1. Amalgamated Dental co. Ltd. 2. Dental Supplies Ltd. 3. S.S. Dental White Ltd.
Wax Modelling Tools	1. Amalgamated Dental Co. Ltd. 2. Ammonite Ltd.

Wax Modelling Tools	3. Chaperlin & Jacobs Ltd.
	4. Hoben Davis Ltd.
Ring Mandrels and Stands	1. Ammonite Ltd.
Wax Modelling Tools (electric)	1. Adcola Ltd.
	2. F. & H. Baxter Ltd.
	3. Hoben Davis Ltd.
	4. W. J. Hooker Ltd.
Wax Injectors	1. Ferraris Engineering & Development Co. Ltd.
	2. Hoben Davis Ltd.
	3. W. J. Hooker Ltd.
	4. Nesor Ltd.
Mould Rubber	1. Hoben Davis Ltd.
	2. W. J. Hooker Ltd.
Electric Vulcanising Press	1. Hoben Davis Ltd.
	2. W. J. Hooker Ltd.
	3. William Frost Products Ltd.
Cold Cure Rubber	1. Alec Tiranti Ltd.
	2. Sasco Ltd.
Stainless Steel Flasks	1. Amalgamated Dental Co. Ltd.
	2. Ammonite Ltd.
	3. Chaperlin & Jacobs Ltd.
	4. Griffin & George Ltd.
	5. S.S. Dental White Ltd.
Jewellery Investment	1. Ammonite Ltd.
	2. Hoben Davis Ltd.
	3. W. J. Hooker Ltd.
Sterling Scientific Investment	1. Engelhard Sales Ltd.
De-foaming Agent	1. Hoben Davis Ltd.
Flexible Mixing Bowls	1. Amalgamated Dental Co. Ltd.
	2. Ammonite Ltd.
Wetting Agent (debubbliser) Trade name Wettax	1. Amalgamated Dental Co. Ltd.
	2. Ammonite Ltd.
	3. Chaperlin & Jacobs Ltd.
	4. S.S. Dental White Ltd.

144

Electromagnetic Vibrators	1. Amalgamated Dental Co. Ltd. 2. Chaperlin & Jacobs Ltd. 3. Metrodent Ltd. 4. Virilium Ltd.
Vacuum Investing Equipment	1. V. N. Barret (Sales) Ltd. 2. Hoben Davis Ltd. 3. W. J. Hooker Ltd.
Rotary Vacuum Pumps	1. V. N. Barrett (Sales) Ltd. 2. Edwards Vacuum Components Ltd. 3. T.W. Associates.
Water Jet Pumps	1. Edwards Vacuum Components Ltd. 2. Griffin & George Ltd.
Investment Flocculation Agents	1. Hoben Davis Ltd. 2. La Port Industries Ltd., General Chemicals Division, Widnes, Lancs.
Burnout Furnaces (electric)	1. Ammonite Ltd. 2. British Ceramic Service Co. Ltd. 3. Carbolite Ltd. 4. Chaperlin & Jacobs Ltd. 5. Hoben Davis Ltd. 6. W. J. Hooker Ltd. 7. Metrodent Ltd.
Burnout Furnaces (gas)	1. Hoben Davis Ltd. 2. W. J. Hooker Ltd. 3. Kasenit Ltd. 4. Wellman Incandescent Ltd.
Sterling Casting Flux for silver and gold	1. Ammonite Ltd. 2. Engelhard Sales Ltd.
Casting Machines Air Pressure	1. Ammonite Ltd. 2. Griffin & George Ltd.

Steam Pressure (Solbrig)	1. Amalgamated Dental Co. Ltd. 2. S.S. Dental White Ltd.
Vertical Centrifuge (spring)	1. Nesor Ltd. 2. Virilium Ltd.
Horizontal Centrifuge (spring)	1. Hoben Davis Ltd. 2. Metrodent Ltd. 3. Nesor Ltd.
Horizontal Centrifuge (motorised)	1. Hoben Davis Ltd. 2. W. J. Hooker Ltd. 3. Metrodent Ltd.
Horizontal Centrifuge (high frequency)	1. Dental Supplies Ltd. 2. Ferraris Engineering & Development Co. Ltd.
Horizontal Centrifuge for low temperature alloys	1. The Foremost Rubber Co. Ltd. 2. N. Saunders Metal Products Ltd.
Vacuum (water pump)	1. Hoben Davis Ltd.
Vacuum (motorised)	1. South West Smelting Co. (U.K. Factor not yet appointed.)
Melting Torches	1. Adaptogas Ltd. 2. Flamefast Engineering Co. Ltd. 3. Sievert Ltd.
Bullion Dealers	1. Ammonite Ltd. (Silver casting grain only.) 2. Engelhard Sales Ltd. 3. Johnson Matthey Metals Ltd. 4. Sheffield Smelting Co. Ltd.
Lead/Tin Alloys	1. Fry's Metals Ltd.
Sand Blasting Machines	1. Chaperlin & Jacobs Ltd. 2. Marcel A. Courtin Ltd. 3. Guyson Industrial Equipment Ltd. 4. Metrodent Ltd.

146

Barrel Polishing Machines	1. W. Canning & Co. Ltd.
	2. Ferraris Engineering & Development Co. Ltd.
Polishing Machines	1. Ammonite Ltd.
	2. Carter Electrical Co. Ltd.
	3. Kernocraft Ltd.
	4. John Quayle Dental Mfg Co. Ltd.
	5. Sewtric Ltd.
Polishing Compounds (including metallic media)	1. W. Canning & Co. Ltd.
Gemstones	1. Ammonite Ltd.
	2. Avon Gems Ltd.
	3. Hillside Gems Ltd.
	4. Kernocraft Rocks & Gems Ltd.
	5. Lapidary Wholesale Supply Co. Ltd.
	6. Levy Gems Co.
	7. Bernard C. Lowe Ltd.
	8. P.M.R. Ltd.
	9. Wessex Impex Ltd.
Synthetic Gems	1. Carmanda
	2. Levy Gems Co.
	3. D. Swarovski & Co. (International) Ltd.
Lapidary Machines	1. Ammonite Ltd.
	2. Avon Gems Ltd.
	3. Griffin & George Ltd.
	4. Hillside Gems Ltd.
	5. Kernocraft Rocks & Gems Ltd.
	6. Lapidary Wholesale Supply Co.
	7. J. Owen Engineering
	8. P.M.R. Ltd.
	9. Wessex Impex Ltd.
Jewellers Tools and Equipment	1. Charles Cooper (Hatton Garden) Ltd.
	2. E. Gray & Son Ltd.
	3. Thomas Sutton (Birmingham) Ltd.
	4. H. S. Walsh & Sons Ltd.

Company addresses

Adcola Ltd.	Adcola House, Gauden Road, London SW4 6LN.
Adaptogas Ltd.	Meadow Mills, Water Street, Stockport, Cheshire. SK1 2BY.
Amalgamated Dental Co. Ltd. (Claudius Ash Ltd.)	Head office: Amalco House, 26–40 Broadwick St., London W1A 2AD. (Branches in most large cities).
Ammonite Ltd.	Llandow Trading Estate, Cowbridge, Glam. CF7 7PB.
T.W. Associates	64 Malthouse Road, Crawley, Sussex, RH10 6BG.
Avon Gems	Strathavon, Boon Street, Eckington, nr Pershore, Worcs.
V. N. Barrett (Sales) Ltd.	1 Mayo Road, Croydon, Surrey. CR0 2QP.
F. & H. Baxter Ltd.	Beckside Lane, Lidget Green, Bradford 7.
Blayson-Olefines Ltd.	Poth Hille Works, 37 Stratford High St., London E15 2QD.
British Ceramic Service Co. Ltd.	Bricesco House, Park Avenue, Wolstanton, Newcastle, Staffs. ST5 8AT.
W. Canning & Co. Ltd.	Great Hampton Street, Birmingham 18.
Carbolite Ltd.	Bamford Mill, Sheffield. S30 2AU.
Carmanda	25 Spring Bank, Hull, HU3 1AF.
Carter Electrical Co. Ltd.	Eastern Avenue, Romford, Essex. RM7 7PD.
Chaperlin & Jacobs Ltd.	591 London Road, North Cheam, Surrey.
Charles Cooper (Hatton Garden) Ltd.	Wall House, 12 Hatton Wall, London EC1.
Marcel A. Courtin Ltd.	4 The Street, Ashtead, Surrey.

Hoben Davis Ltd.	Spencroft Road, Newcastle, Staffs.
Dental Supplies Ltd.	11 All Saints Road, London W11.
Alexander Duckham & Co. Ltd.	Rainville Road, Hammersmith, London, W6 9HA
Edwards Vacuum Components Ltd.	Manor Royal, Crawley, Sussex. RH.10 2LW.
Engelhard Sales Ltd.	49/63 Spencer Street, Birmingham B18 6DQ.
Ferraris Engineering & Development Co. Ltd.	26 Lea Valley Trading Estate, Angel Road, Edmonton, London N18 3JD.
Flamefast Engineering Ltd.	Pendlebury Industrial Estate, Bridge Street, Swinton, Manchester M27 1FJ.
The Foremost Rubber Co. Ltd.	22 Hollen St., London W1V 4BX.
W. Frost Products Ltd.	Amberly Way, Green Lane, Hounslow, Middx.
Fry's Metals Ltd.	Tandem Works, Merton Abbey, London SW19.
E. Gray & Son Ltd.	Grayson House, 12–16 Clerkenwell Road, London EC1.
Griffin & George Ltd. Technical Studies	P.O. Box 14, Wembley, Middx. HA0 1HJ.
Guyson Industrial Equipment Ltd.	North Avenue, Otley, Yorks. LS21 1AR.
Hillside Gems Ltd.	1 Florence Road, Wylde Green, Sutton Coldfield, Warwicks.
W. J. Hooker Ltd.	Water Side, Brightlingsea, Colchester, Essex.
Kasenit Ltd.	Denbigh Road, Bletchley, Bucks.
Kernocraft Rocks & Gems Ltd.	44 Lemon St., Truro, Cornwall.
Lapidary Wholesale Supply Co.	44 Walmsley St., Hull. HU3 1QD.
Levy Gems Co.	92 Hatton Garden, London EC1.
Bernard C. Lowe	73–75 Spencer Street, Birmingham 18.

Johnson Matthey Metals Ltd.	Vittoria St., Birmingham.
Metrodent Ltd.	P.O. Box 829, 15 Chancery Lane, Huddersfield HD1 2DU.
Nesor Products Ltd.	Claremont Hall, Pentonville Rd., The Angel, London N1.
J. Owen Engineering	14 Downs Road, Maldon, Essex.
P.M.R. Ltd.	Smithy House, Atholl Road, Pitlochry, Perthshire, Scotland.
John Quayle Dental Mfg. Co. Ltd.	19 Buckingham Road, Worthing, Sussex.
Sasco Ltd.	P.O. Box 20, Gatwick Road, Crawley, Sussex.
N. Saunders Metal Products Ltd.	Enessa Works, Edwin Road, Twickenham, Middx.
Sew-tric Ltd.	Sewtric House, Honeypot Lane, Stanmore, Middx. HAZ 1J2.
Sheffield Smelting Co.	Royds Mill St., Sheffield 4.
Sievert	Wm. A. Mayer Ltd., 9–11 Glendelon Road, London SW16.
South West Smelting & Refining Co.	1712 Jackson St., P.O. Box 210, Dallas, Texas 75221.
S.S. Dental White Co.	Nechells House, Dartmouth St., Birmingham 7.
Thomas Sutton (Birmingham) Ltd.	37 Frederick Street, Birmingham 1.
D. Swarovski Co. International Ltd.	43–51 Great Titchfield St., London W1P 7FJ.
Tiranti Ltd.	72 Charlotte St., London W1.
The Virilium Co. Ltd.	46–48 Pentonville Road, London N1.
H. S. Walsh & Sons Ltd.	243 Beckenham Road, Beckenham, Kent. BR3 4RP.
Wellman Incandescent Furnace Co. Ltd.	Cornwall Road, Smethwick, Warley, Worcs.
Wessex Impex Ltd.	Gemini, Lanham Lane, Winchester, Hants.

Candle Makers Supplies, of 4 Beaconsfield Terrace Road, London W14, can now offer a comprehensive range of materials and equipment for the lost wax casting of jewellery.

The following American companies supply casting equipment and materials and will send illustrated catalogues on request:

Allcraft Tool & Supply Co. Inc.	215 Park Avenue, Hicksville, New York 11801.
Kerr Manufacturing Co.	28200 Wick Road, Romulus, Michigan 48174.
Romanoff Rubber Company, Inc.	153–159 West 27th Street, New York, N.Y. 10001.
South West Smelting & Refining Co.	P.O. Box 2010, 1712 Jackson, Dallas, Texas 75221.
Technical Specialities International Inc.	487 Elliot Avenue West, Seattle, Washington 98119.

Appendix III. Conversion tables

WEIGHT

To Convert:

Grains to grammes	multiply by	.0647989
Grammes to grains	"	15.4324
Pennyweights to grammes	"	1.55518
Grammes to pennyweights	"	.64301
Ounces troy to grammes	"	31.1035
Grammes to ounces troy	"	.0321507
Ounces avoirdupois to grammes	"	28.3495
Grammes to ounces avoirdupois	"	.0352740
Ounces avoirdupois to grains	"	437.5
Grains to ounces avoirdupois	"	.0022857
Ounces troy to grains	"	480.0
Grains to ounces troy	"	.0020833
Ounces troy to ounces avoirdupois	"	1.09714
Ounces avoirdupois to ounces troy	"	.911458
Ounces troy to pounds avoirdupois	"	.06857
Pounds avoirdupois to ounces troy	"	14.583328
Pounds avoirdupois to kilograms	"	.4535924
Kilograms to pounds avoirdupois	"	2.20462
Pounds avoirdupois to grains	"	7000.0
Grains to pounds avoirdupois	"	.0001428
Kilograms to ounces avoirdupois	"	35.2740
Kilograms to ounces troy	"	32.1507

LENGTH

To Convert:

Millimetres to inches	multiply by	.0393701
Inches to millimetres	"	25.4
Centimetres to inches	"	.393701
Inches to centimetres	"	2.54
Metres to inches	"	39.3701
Inches to metres	"	.0254
Feet to metres	"	.3048
Metres to feet	"	3.28084
Yards to metres	"	.9144
Metres to yards	"	1.09361

AREA AND VOLUME

To Convert:

Square inches to square millimetres	. . . . multiply by	645.16
Square inches to square centimetres	. . . . „	6.4516
Square centimetres to square inches	. . . . „	.1550
Square millimetres to square inches	. . . . „	.00155
Cubic inches to cubic centimetres .	. . . . „	16.3871
Cubic centimetres to cubic inches .	. . . . „	.061024

TEMPERATURE

To Convert:

°F to °C   Subtract 32, multiply by 5, and divide by 9
°C to °F   Multiply by 9, divide by 5, and add 32

Decimal and metric equivalents of common fractions

Fractions of an inch	Decimals of an inch	Equivalent in millimetres	Fractions of an inch	Decimals of an inch	Equivalent in millimetres
1/64	.01562	.397	33/64	.51562	13.097
1/32	.03125	.794	17/32	.53125	13.494
3/64	.04687	1.191	35/64	.54687	13.891
1/16	.0625	1.588	9/16	.5625	14.288
5/64	.07812	1.984	37/64	.57812	14.684
3/32	.09375	2.381	19/32	.59375	15.081
7/64	.10937	2.778	39/64	.60937	15.478
1/8	.1250	3.175	5/8	.625	15.875
9/64	.14062	3.572	41/64	.64062	16.272
5/32	.15625	3.969	21/32	.65625	16.669
11/64	.17187	4.366	43/64	.67187	17.066
3/16	.1875	4.763	11/16	.6875	17.463
13/64	.20312	5.159	45/64	.70312	17.859
7/32	.21875	5.556	23/32	.71875	18.256
15/64	.23437	5.953	47/64	.73437	18.653
1/4	.2500	6.350	3/4	.75	19.050
17/64	.26562	6.747	49/64	.76562	19.447
9/32	.28125	7.144	25/32	.78125	19.844
19/64	.29687	7.541	51/64	.79687	20.241
5/16	.3125	7.938	13/16	.8125	20.638
21/64	.32812	8.334	53/64	.82812	21.034
11/32	.34375	8.731	27/32	.84375	21.431
23/64	.35937	9.128	55/64	.85937	21.828
3/8	.3750	9.525	7/8	.875	22.225
25/64	.39062	9.922	57/64	.89062	22.622
13/32	.40625	10.319	29/32	.90625	23.019
27/64	.42187	10.716	59/64	.92187	23.416
7/16	.4375	11.113	15/16	.9375	23.813
29/64	.45312	11.509	61/64	.95312	24.209
15/32	.46875	11.906	31/32	.96875	24.606
31/64	.48437	12.303	63/64	.98437	25.003
1/2	.5	12.700	1	1.000	25.400

Comparative weights—for use with precious metals

TROY WEIGHT—now superceded by the metric system
24 grains = 1 pennyweight (dwt)
20 dwt = 1 troy ounce
12 oz   = 1 pound troy

GRAM WEIGHTS
    1 gram = 15.43 grains troy
  1.555 grams = 1 dwt
31.104 grams = 1 ounce troy
28.35 grams = 1 ounce avoirdupois

Carat weights—for use with precious and semi-precious gemstones
1 carat = $3\frac{1}{16}$ grains troy
1 „ = .007 ounce avoirdupois
1 „ = .20 grams

The carat is further divided as follows:

1 carat = 100 points
$\frac{1}{2}$ „ = 50/100 points
$\frac{1}{4}$ „ = 25/100 points
$\frac{1}{8}$ „ = $12\frac{1}{2}$/100 points

Appendix IV. Reagents

Pickling solutions

1. Sulphuric acid   2 parts   Suitable for silver, gold, brass, copper, bronze,
   Water   10 parts   and nickel silver. A small amount of potassium dichromate added to the solution may improve the action when pickling gold.

2. Safe-T-Pickle   one rounded   This is a sulphuric acid compound supplied
      table spoon   in granular form. Though its action may be
   Water   one half litre   a little slower than the previous pickle, it is recommended for schools because of ease of preparation.

3. Hydrochloric acid   1 part   Much more rapid in its action when removing
   Water   3 parts   oxides and glazed investment deposits. Because of the corrosive action of its fumes, it should be used in a well ventilated area. Suitable for silver and gold.

4. Hydrofluoric acid   1 part   Used for rapid removal of oxides and glazed
   Water   3 parts   investment from gold and platinum. **Its toxic fumes and corrosive nature make it highly dangerous if handled carelessly.** This acid will attack glass and should be stored in lead, plastic or wax containers.

Neutralising agent

Carbonate of soda   100 grams   As all the above acids belong to the same
Water   4 litres   group, a common neutralising solution may be used and will be more effective if slightly warmed.

Appendix V.
Precious metals

Availability of precious metals

Whilst the common forms of silver may be obtained 'over the counter' at most bullion dealers, the purchase of the various gold alloys presents more of a problem for the amateur.

Current Bank of England regulations stipulate that a Bullion Licence application form must first be completed and certified before gold may be purchased. The form, known as a G.A. certificate, may be obtained from any recognised bullion dealer and is usually lodged with the dealer when official approval has been granted.

A word of warning though: Official sanction is unlikely unless the applicant is a manufacturing jeweller, or teaches in an educational establishment which runs full-time courses in jewellery. In the latter case, the school or college would be granted the licence.

Johnson Matthey Assayable Silver and Gold Solders

HALL-MARKING QUALITY SILVER SOLDERS

Description	Melting Range °C	Characteristics	Available Sizes mm	[]	in	Recommended Flux
Extra Easy	667–709 }	For general soldering giving strong ductile joints	2.0 × 0.45		.079 × .018	Easy-flo
Easy*	705–723 }		3.0 × 0.50		.118 × .020	Easy-flo
			50.0 × 0.50		1.97 × .020	
			50.0 × 0.25		1.97 × .010	
			Wire 0.5 dia.		.020 dia.	
Medium	720–765	Also a general purpose solder, of higher silver content than Easy	1.5 × 0.7		.059 × .028	Tenacity Flux No. 5
			75.0 × 0.7		2.95 × .028	
Hard	745–778	For use in two-stage soldering where a second joint is to be made with Easy solder	5.0 × 0.6		.197 × .024	Tenacity Flux No. 5
			75.0 × 0.6		2.95 × .024	
Enamelling	730–800	For work that has subsequently to be enamelled, and also for first soldering operations	1.5 × 1.0		.059 × .039	Tenacity Flux No. 5

* Easy solder is also available in powder form.

GOLD SOLDERS—CHARACTERISTICS AND AVAILABILITY

Solder	Gold Assay parts per 1000	Melting Range °C	2 g pieces	Availability * Ex Stock † Made to order sheet 50 mm × 0.5 mm	wire	Recommended Flux
YELLOW						
9 carat Extra Easy	375	640–652	*	*	*	Easy-flo
9 carat Easy‡	375	695–715	*	*	*	Easy-flo
9 carat Medium	375	720–760	*	*	†	Easy-flo
9 carat Hard	375	756–793	*	*	†	Tenacity No. 5
14 carat Easy	585	703–730	*	*	†	Easy-flo
14 carat Hard	585	753–783	*	*	†	Tenacity No. 5
15 carat Easy	625	744–765	*	†	†	Tenacity No. 5
18 carat Easy‡	750	633–705	*	*	†	Easy-flo
18 carat Medium	750	739–752	*	*	†	Easy-flo
18 carat Hard	750	815–825	*	*	†	Tenacity No. 5
EYG 800	800	866–877	*	†		Tenacity No. 5
RED						
9 carat Red	375	734–784	*	†	†	Tenacity No. 5
18 carat Red	750	796–819	*	†	†	Tenacity No. 5
WHITE						
9 carat White	375	710–743	*	†	†	Easy-flo
EWG 496	496	690–711	*	†		Easy-flo
MWG 588	588	703–720	*	†		Easy-flo
HWG 833	833	853–885	*	†		Tenacity No. 5

‡ Also available in powder form to 90, 120, 200 mesh.

Carat golds

Characteristics

Alloy	Colour	Melting range °C Solidus Liquidus	Maximum hardness of annealed sheet Hv	Minimum elongation of annealed sheet 50 mm gauge per cent	An
9 carat P	Pale Yellow	905–960	170	30	
9 carat BY	Pale Yellow	890–920	145	20	
9 carat DF	Yellow	880–900	120	40	
9 carat C	Yellow	885–895	120	45	
9 carat SC	Yellow	875–890	110	60	
9 carat G	Green	800–820	100	65	
9 carat BR	Red	890–915	140	30	
9 carat MR	Deep Red	900–920	110	35	
9 carat HW	White	975–1025	160	35	
9 carat MW	White	910–940	90	40	
9 carat SW	White	990–1005	45	40	
14 carat HB	Pale Yellow	930–965	95	35	
14 carat JP	Pale Yellow	835–865	160	25	
14 carat AF	Yellow	830–875	130	50	
14 carat WA	Yellow	830–845	180	58	
14 carat M	Yellow	820–860	150	45	
14 carat DR	Red	855–870	180	40	
14 carat MW	White	970–985	150	45	
14 carat SW	White	1170–1250	85	30	

Applications   Availability

R = Recommended
S = Suitable

* some sizes stocked
† made to order
‡ not normally made

...ling ...hod / ...ing note ...ow	A – General purpose	B – Spinning	C – High relief stamping	D – Deep drawing	E – Enamelling	F – Chain making	G – Pen nib manufacture	H – Investment casting	I – Snaps, pins, springs	Sheet	Wire	Tube	Alloy
1								S	R	*	*	‡	9 carat P
3				R				S		†	†	‡	9 carat BY
3	R		R		R		R	S		*	*	*	9 carat DF
3	S	S			R					†	†	‡	9 carat C
3	S	S			R		S			†	†	‡	9 carat SC
3	S	R								*	*	‡	9 carat G
1				R			S			†	†	‡	9 carat BR
3	S		R			S	S			*	*	†	9 carat MR
2	R			R		S	S	R		*	*	†	9 carat HW
3	R		S	S	R		R			*	*	†	9 carat MW
3		R								*	*	‡	9 carat SW
3	R	R	S	S	R	R		R		*	*	†	14 carat HB
1	R		R	R	R		R	R		*	*	†	14 carat JP
3	R		R		R		R			†	†	†	14 carat AF
1						R				†	†	‡	14 carat WA
3	R		R		R	R	R	S		*	*	†	14 carat M
1						R				†	‡	‡	14 carat DR
2	R		R				R	R		†	*	†	14 carat MW
3	S			R						†	†	‡	14 carat SW

Method of cooling:
1 *Must* be quenched from above 500°C
2 *Must not* be quenched
3 *May* be quenched once the metal has cooled to black heat (450°–500°C)
Alloys marked 1 are age hardenable.

Carat golds (continued)

Characteristics

Alloy	Colour	Melting range °C Solidus Liquidus	Maximum hardness of annealed sheet Hv	Minimum elongation of annealed sheet 50 mm gauge per cent	Ann Te
18 carat HB	Yellow	895–930	125	40	6
18 carat FG	Green	960–1025	45	35	6
18 carat AK	Red	855–870	155	45	6
18 carat MR	Deep Red	875–900	155	45	6
18 carat FW	White	890–960	230	35	7
18 carat MW	White	1180–1235	130	25	7
18 carat SW	White	1300–1315	70	30	6
18 carat CW	White	1090–1100	160		7
22 carat DS	Yellow	965–980	75	30	6
22 carat R	Red	930–950	90	30	6

Applications   Availability

R = Recommended
S = Suitable

* some sizes stocked
† made to order
‡ not normally made

A – General purpose	B – Spinning	C – High relief stamping	D – Deep drawing	E – Enamelling	F – Chain making	G – Pen nib manufacture	H – Investment casting	I – Snaps, pins, springs	Sheet	Wire	Tube	Alloy
R	R		R	R	R		R		*	*	*	18 carat HB
	R								*	*	‡	18 carat FG
R						S	S		†	†	†	18 carat AK
S							S		*	*	†	18 carat MR
R		R			R	S	R		*	*	†	18 carat FW
R		R	S	S					*	*	*	18 carat MW
S			R						*	*	‡	18 carat SW
							R		†	‡	‡	18 carat CW
R	R	S	R	R			R		*	*	†	22 carat DS
	R						R		*	*	†	22 carat R

Method of cooling:

Must be quenched from above 500°C
Must not be quenched
May be quenched once the metal has cooled to black heat (450°–500°C)

Alloys marked **1** are age hardenable.

Engelhard carat gold solders and precious metal alloys

ENGELHARD ALLOY	ENGELHARD CODE No.	PRINCIPAL APPLICATIONS
97% Platinum	97	Claw settings, brooches, bracelets, rings, etc.
97% Platinum 126	126	Delicate stone setting.
Platinum/Iridium (containing 5% to 25% Iridium)	56 & others	Hard alloys for safety catches, pins, springs, watch case backs, engine turning, cigarette cases, etc. (Higher Iridium content indicates harder alloy.)
839 Palladium	839	As for Platinum 97 alloy. Will not tarnish.
CARAT GOLDS 22 ct Yellow 22 ct Red	512 636	General purpose alloys for wedding rings and fine jewellery products where hard wearing properties are not a prime consideration.
18 ct White A	461	Hard gold for fine wire, galleries, etc.
18 ct White D	466	Casting gold for large stones and cluster settings.
18 ct White G	462	Settings.
18 ct White R	469	Ring and Shanking wires.
18 ct White C	473	Casting gold. Very white appearance.
18 ct White 470	470	Casting gold. Containing Platinum.
18 ct White 475	475	Hard casting gold for settings, etc.
18 ct White 476	476	Soft casting gold for settings, etc.
18 ct Yellow B	534	General jewellery purposes.
18 ct Green	626	General jewellery purposes.
18 ct Red B	647	General jewellery purposes.
18 ct Med Red	646	Casting alloy
9 ct White K	505	Soft alloy. Very workable. Eternity rings. Tubing.
9 ct White M	506	Medium gold. Eternity rings. Brooches, etc.
9 ct White C	503	Casting gold. Ideal for cast eternity rings.
9 ct Yellow DF	610	General purpose jewellery alloy.
9 ct Yellow M	614	Paler than DF. Casting alloy.
9 ct Green M	633	General jewellery product purposes.
9 ct Red B	692	General jewellery product purposes.
Enamelling	617	Enamelling.
Britannia Silver	1365	General purpose
Sterling Silver	1375	General purposes. Castings, sheet, metal work, etc.
Sterling Silver	1376	Very free cutting.
14 ct White A	480	These alloys comply
14 ct Yellow B	556	with foreign
14 ct White F	483	specification.
14 ct White E	485	
10 ct White M	495	
10 ct Red	687	
10 ct Yellow E	590	

STANDARD FORMS AVAILABLE FOR THE MANUFACTURING JEWELLER

Findings	Sheet	Strip	Wire	Eternity Ring Wire	Grain	Wedding Ring Tube	Seamless Tube	Castings	Wedding Ring Cut Pieces
●	●	●	●	●			●	●	
	●								
			●						
●	●	●	●	●	●		●	●	
	●	●	●			●	●		●
	●	●	●			●	●		●
	●		●						
	●	●	●	●	●		●	●	
●	●	●	●			●	●		
	●	●	●						
					●			●	
					●			●	
					●			●	
					●			●	
●	●	●	●	●	●	●	●	●	●
	●	●	●				●		
	●	●	●				●		
					●			●	
	●	●	●	●		●	●		
	●	●	●	●			●		
					●			●	
●	●	●	●	●	●	●	●	●	●
					●			●	
	●	●	●				●		
	●	●	●				●		
	●	●							
	●	●	●						
	●	●	●		●		●	●	
	●	●	●		●		●	●	
	●		●		●				
	●	●	●		●			●	
	●		●		●				
	●		●		●				
	●		●		●				
	●		●		●				
	●		●		●				
	●		●		●				

Alloy	Engelhard Code No.	Solidus °C	Liquidus °C	Hardness Annealed HV	Annealing Temperature °C	Conversion Factor for use with Platinum Weight Tables
97% Platinum	97	1740	1750	55	1000	0.96
97% Platinum	126	1740	1750	55	1000	0.96
Platinum/Iridium	54	1800	1810	160	1050	1.01
839 Palladium	839	1579	1645	90	1000	0.56
Carat Golds						
22ct yellow	512	983	1002	74	450	0.83
22ct red	636	957	978	89	450	0.03
18ct white A	461	900	988	249	800	0.68
18ct white D	466	1265	1338	105	550	0.77
18ct white G	462	1008	1085	183	650	0.74
18ct white R	469	1124	1205	140	800	0.75
18ct white C	473	888	967	238	650	0.68
18ct white 470	470	1188	1278	70	800	0.79
18ct white 475	475	893	926	209	700	0.70
18ct white 476	476	948	1010	150	650	0.73
18ct yellow B	534	898	909	138	700	0.72
18ct green 626	626	938	975	82	800	0.74
18ct red B	647	897	903	159	700	0.70
9ct white K	505	874	942	74	550	0.56
9ct white M	506	982	1071	125	800	0.51
9ct white C	503	919	963	60	650	0.59
9ct yellow DF	610	825	902	125	650	0.52
9ct yellow M	614	809	903	120	650	0.52
9ct green M	633	789	863	85	650	0.51
9ct red B	692	990	1003	102	750	0.52
9ct yellow enamelling	617	779	820	155	650	0.56
14ct white A	480					0.59
14ct yellow B	556	833	859			0.63
14ct white F	483				700	0.68
14ct white E	485					0.69
10ct white M	495					0.53
10ct red 687	687					0.53
10ct yellow E	590				650	0.54
Sterling Silver	1375		895	60	500	0.48
Sterling Silver + Tellurium	1376		895	60	500	0.48

Casting grain

ENGELHARD ALLOY NO.		
466	18 ct WHITE GOLD D	A high fusing investment; oxy/gas flame, very malleable. Ideal as setting for large stones and clusters.
476	18 ct WHITE GOLD	A high fusing investment; oxy/gas flame. Medium hardness, direct setting.
475	18 ct WHITE GOLD	A low fusing investment; gas/air flame. Slightly harder than 476. Direct setting and illusion disc setting.
473	18 ct WHITE GOLD C	A low fusing investment; gas/air flame. Hard and very white in appearance. Disc setting.
534	18 ct YELLOW GOLD B	A low fusing investment; gas/air flame. Medium hard. Suitable for all jewellery applications in this carat.
646	18 ct MEDIUM RED	A low fusing investment; gas/air flame. Hard. Suitable for all jewellery applications in this carat.
556	14 ct. YELLOW GOLD B	A low fusing investment; gas/air flame. Suitable for all jewellery applications.
503	9 ct WHITE GOLD C	A low fusing investment; gas/air flame. Medium hard. Very malleable. Ideally suitable for eternity rings.
610	9 ct YELLOW GOLD DF	A low fusing investment; gas/air flame. Medium hard. Excellent general purpose jewellery medium.
614	9 ct YELLOW GOLD M	A low fusing investment; gas/air flame. Very malleable. Paler colour than DF. Ideal eternity ring material. Also brooch production and charms, etc.
1375	STERLING SILVER	A low fusing investment; gas/air flame. Ideal for rings, brooches, charms, etc.
97	PLATINUM 97%	A high fusing investment; oxy/gas flame or induction. Ideal for cluster, stone rings, etc. Malleable but not too soft.

Carat gold solders

CAT. REF. NO.	ALLOY NO.	MELTING POINT APPROX. °C.	FOR USE WITH
J/180	780	HIGH	9 ct YELLOW*
J/181	781	MEDIUM	9 ct YELLOW*
J/182	782	LOW	9 ct YELLOW*
J/183	743	HIGH	9 ct WHITE*
J/184	770	HIGH	14 ct YELLOW*
J/185	733	HIGH	14 ct WHITE*
J/186	755	HIGH	18 ct YELLOW*
J/187	752	MEDIUM	18 ct YELLOW*
J/188	754	LOW	18 ct YELLOW*
J/189	734	MEDIUM	14 ct WHITE
J/190	735	LOW	14 ct WHITE
J/191	727	MEDIUM	18 ct WHITE
J/192	728	LOW	18 ct WHITE
J/193	801		22 ct RED

*HALL MARKING QUALITIES

Specific gravities and melting temperatures of some casting metals

Metal	Melting temperature (liquidus) °C	Specific gravity
Aluminium	700	2.7
Brass	930	8.5
Bronze	1000	9.0
Standard silver (sterling)	890	10.3
9ct gold yellow	880	11.2
9ct gold white	940	12.5
14ct gold yellow	860	13.1
14ct gold white	990	14.9
18ct gold yellow	910	15.5
18ct gold white	950	14.7
22ct gold yellow	1000	17.7
Palladium (pure)	1550	12.0
Platinum	1772	21.5

Metal temperatures

The following table gives a fairly accurate assessment of temperature when metal is heated under subdued lighting.

	°C
Faint red	343–371
Dull red	371–482
Medium red	482–693
Bright red	693–865
Cherry red	871–954
Orange	982–1010
Bright yellow	1023–1052
Yellow/white	1065–1145
White	1191–1242
Blinding white	1353 and higher

Appendix VI.
Assay and hallmarking

A piece of gold or silver jewellery will gain intrinsically in value if it is hallmarked by the Assay Office. On the other hand it is an offence, for example, to offer a piece for sale purporting to be 9ct gold if it does not contain the correct proportion of gold for that particular alloy.

No matter how great the temptation, it would be most unwise to add a different metal to the melt in order to make up the required weight for a given casting. The exacting tests carried out during assay would inevitably reveal the 'fraud' and the Assay Office reserve the right to destroy any item which does not meet the minimum standard for the stated alloy. Similarly, all hard solders used on a piece should be of assayable quality.

Only a fraction of one per cent are usually destroyed but it is significant to note that of this proportion about 65 per cent are investment castings.

This form of consumer protection has been in continuous use since the thirteenth century and the following list of statutory offences may be of cautionary interest:

Table of Offences
under the
Statutes Relating to Gold and Silver Wares.

Manufacturing and trading, without registering at Assay Office, Name, Mark, and Place of Abode.

Selling, exchanging or exposing to sale a Ware before it is marked.

Marking a Ware with a Mark other than that registered at Assay Office.

Introducing or concealing iron or other base metal in a Ware.

Fraudulently erasing, obliterating, or defacing an Assay Mark or registered trader's Mark.

Altering the Character or Denomination of a Ware which has been assayed and marked, or making any addition to it, without first obtaining the assent of the Assay Office to such alteration or addition.

Selling, exchanging or having possession of a Ware unlawfully altered or added to.

The following offences are FELONIES, and punishable by imprisonment:

Forging or counterfeiting any Die or other Instrument used at Assay Office.

Marking Wares with a Forged Die or other Instrument.

Counterfeiting Assay Marks.

Transposing Assay Marks from one Ware to any other.

Having possession, without lawful excuse, of a Forged Die or other Instrument, or of a Ware marked with a Forged Die or other Instrument.

Cutting from a Ware an Assay Mark, with intent to affix it to another Ware.

Using, with intent to defraud, a Die or other Instrument of the Assay Office.

Having possession, knowingly and without lawful excuse or authority, of a Ware bearing a Counterfeit or Transposed Mark.

Knowingly and wilfully aiding or abetting the commission of these offences.

This table has been reproduced by kind permission of the Birmingham Assay Office.

Before one can have work hallmarked, it is necessary to register with the Assay Office and a punch bearing the initials or the trade mark of the applicant must be submitted to the Assay Master for approval.

For a small fee the Assay Office will supply punches of a size consistent with the work to be undertaken. The choice of punch impression may be chosen from many illustrated on the chart provided.

A cranked punch for rings and a straight punch for brooches and the like will fulfil most amateur requirements. Each punch is stamped with the user's registration number and this number must be indicated on the application form sent with the pieces for hallmarking. A completed sample application form is illustrated in Fig. 202.

Fig. 202. A completed application form for assay.

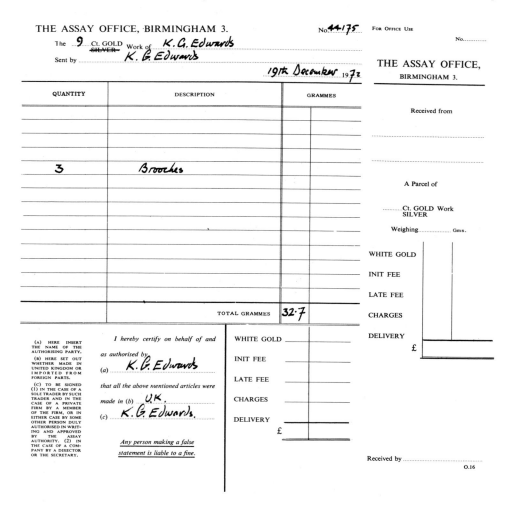

THE ASSAY OFFICE, BIRMINGHAM 3.   No. 44175

The *9* Ct. GOLD ~~SILVER~~ Work of *K. G. Edwards*

Sent by *K. G. Edwards*

19th December 19*72*

QUANTITY	DESCRIPTION	GRAMMES
3	*Brooches*	
	TOTAL GRAMMES	*32.7*

(A) HERE INSERT THE NAME OF THE AUTHORISING PARTY.
(B) HERE SET OUT WHETHER MADE IN UNITED KINGDOM OR IMPORTED FROM FOREIGN PARTS.
(C) TO BE SIGNED (1) IN THE CASE OF A SOLE TRADER BY SUCH TRADER AND IN THE CASE OF A PRIVATE FIRM BY A MEMBER OF THE FIRM, OR IN EITHER CASE BY SOME OTHER PERSON DULY AUTHORISED IN WRITING AND APPROVED BY THE ASSAY AUTHORITY. (2) IN THE CASE OF A COMPANY BY A DIRECTOR OR THE SECRETARY.

I hereby certify on behalf of and
as authorised by
(a) *K. G. Edwards*
that all the above mentioned articles were
made in (b) *U.K.*
(c) *K. G. Edwards*

Any person making a false statement is liable to a fine.

WHITE GOLD	
INIT FEE	
LATE FEE	
CHARGES	
DELIVERY	
£	

FOR OFFICE USE   No. ....

THE ASSAY OFFICE,
BIRMINGHAM 3.

Received from
....
....

A Parcel of
............Ct. GOLD Work
          SILVER
Weighing.................Gms.

WHITE GOLD
INIT FEE
LATE FEE
CHARGES
DELIVERY
          £

Received by ....
                              O.16

170

The various application forms and a most interesting booklet describing the work of the Assay Office may be obtained on request from:

The Assay Master,
The Assay Office,
Newhall Street, Birmingham 3.

Minimum percentages of Gold and Silver for Hall-marking purposes

The quality of gold is denoted by carat (ct) amounts. Pure gold is 24ct and is for the most part too soft for jewellery.

The addition of other metal elements such as copper, silver, zinc, and nickel imparts certain desirable working qualities or colours, e.g. white, red, and green, to the resulting alloy.

22ct gold	91.66% or 22/24ths
18ct gold	75% or 18/24ths
14ct gold	58.5% or 14/24ths
9ct gold	37.5% or 9/24ths.

Similarly with silver alloys.

Sterling Silver   is 92.5% pure silver
Britannia Silver is 95.8% pure silver.

A hallmark is made up of several symbols:

1. The maker's mark consisting of the initials of the person or company submitting the piece for assay.
2. The standard mark denoting the minimum gold or silver content.
3. The Assay Office mark showing which Office tested the piece.
4. The date letter which indicates the year in which the piece was hallmarked.

Fig. 203. A typical hall-mark.

There were formerly Assay Offices in many cities, each having its own distinctive mark, but possibly due to the regionalisation of the jewellery trade, there are currently only four in the U.K., situated in London, Birmingham, Sheffield, and Edinburgh.

Fig. 204 illustrates the marks of the four Assay Offices, together with the standard marks for gold and silver.

Fig. 204.

Mark	Standard	Minimum percentage
	Sterling silver Marked in England	92.5
	Sterling silver Marked in Scotland	92.5
	Britannia silver	95.84
 22	22 carat gold Marked in England	91.66
 18	18 carat gold Marked in England	75.0
 14 585	14 carat gold	58.5
 9 375	9 carat gold	37.5

	London	Sterling silver & gold
	London	Britannia silver
	Birmingham	silver & gold
	Sheffield	silver
	Sheffield	gold
	Edinburgh	silver & gold

Appendix VII.
Principal gems of the world and their properties

Chemical Formula	Name of Gem	System of Crystal-lography *	Mohs's Hard-ness	Specific Gravity	Refrac-tive Index	Colour
SiO_2	AGATE	Crypto-Cry.	7·0	2·62	1·53	Various stratifie
$BeAl_2O_4$ or $BeO. Al_2O_3$	ALEXANDRITE	Rhombic	8·5	3·72	1·74–1·75	G. Daylight. R Ar
$KAlSi_3O_8$ or $K_2O. Al_2O_3. 6SiO_2$	AMAZONSTONE	Triclinic	6·5	2·57	1·52–1·53	G
SiO_2	AMETHYST	Trigonal	7·0	2·65	1·54–1·55	M
$Be_3Al_2Si_6O_{18}$ or $3BeO. Al_2O_3. 6SiO_2$	AQUAMARINE	Hexagonal	7·5	2·70	1·57–1·58	B G
SiO_2	AVENTURINE QUARTZ	Trigonal	7·0	2·65	1·54–1·55	G R &
$Be_3Al_2Si_6O_{18}$ or $3BeO. Al_2O_3. 6SiO_2$	BERYL	Hexagonal	7·5	2·70	1·57–1·58	G Y B W &
SiO_2	BLOODSTONE	Crypto-Cry.	7·0	2·62	1·53	G splashe
$H_4Mg_3Si_2O_9$	BOWENITE	Monoclinic	5·0	2·58	1·56	G
SiO_2	CAIRNGORM	Trigonal	7·0	2·65	1·54–1·55	Y-Br
SiO_2	CORNELIAN	Crypto-Cry.	7·0	2·62	1·53	Y & R
$BeAl_2O_4$ or $BeO. Al_2O_3$	CAT'S EYE—CHRYSOBERYL	Rhombic	8·5	3·72	1·74–1·75	Y-G with
SiO_2	CAT'S EYE—QUARTZ	Trigonal	7·0	2·65	1·54–1·55	Gy W & Br w
SiO_2	CHALCEDONY	Crypto-Cry.	7·0	2·62	1·53	G, B &
$BeAl_2O_4$ or $BeO. Al_2O_3$	CHRYSOBERYL	Rhombic	8·5	3·72	1·74–1·75	Y G &
SiO_2 (Nickel as impurity)	CHRYSOPRASE	Crypto-Cry.	7·0	2·62	1·53	G
SiO_2	CITRINE	Trigonal	7·0	2·65	1·54–1·55	Y
Al_2O_3	CORUNDUM	Trigonal	9·0	3·99	1·76–1·77	W Y R B M
Al_2O_3	CORUNDUM (Syn)	Trigonal	9·0	3·99	1·76–1·77	W Y R B M
C	DIAMOND	Cubic	10·0	3·52	2·42	W B G Br Pk
Cu minerals	EILAT STONE	Rock	4·0	2·8–3·2	—	Variable G, B
$Be_3Al_2Si_6O_{18}$ or $3BeO. Al_2O_3. 6SiO_2$	EMERALD	Hexagonal	7·5	2·69	1·57–1·58	G
$Be_3Al_2Si_6O_{18}$ or $3BeO. Al_2O_3. 6SiO_2$	EMERALD (Syn)	Hexagonal	7·5	2·65	1·56	G
CaF_2	FLUORSPAR	Cubic	4·0	3·18	1·43	W B M Pk Y
$Fe_3Al_2(SiO_4)_3$	GARNET—ALMANDINE	Cubic	7·5	4·20	1·79	M-R
$Ca_3Fe_2(SiO_4)_3$	GARNET—ANDRADITE	Cubic	6·5	3·85	1·88	Bk
$Ca_3Fe_2(SiO_4)_3$	GARNET—DEMANTOID	Cubic	6·5	3·84	1·88	G
$Ca_3Al_2(SiO_4)_3$	GARNET—GROSSULARITE	Cubic	7·25	3·63	1·74	G
$Ca_3Al_2(SiO_4)_3 + OH$	GARNET—GROSSULARITE	Massive	7·0	3·45	1·72	G Pk R
$Ca_3Al_2(SiO_4)_3$	GARNET—HESSONITE	Cubic	7·25	3·65	1·74	Br
$Mg_3Al_2(SiO_4)_3$	GARNET—PYROPE	Cubic	7·25	3·75	1·75	R
$(Fe, Mg)_3Al_2(SiO_4)_3$	GARNET—RHODOLITE	Cubic	7·50	3·84	1·76	R-M
$Mn_3Al_2(SiO_4)_3$	GARNET—SPESSARTITE	Cubic	7·25	4·20	1·81	Br-Y
A mixture of silicates	GOLDSTONE	—	5·5	2·60	1·53	B & Br
Fe_2O_3	HÆMATITE	Trigonal	6·0	5·10	—	Bk
$LiAl(SiO_3)_2$ or $Li_2O. Al_2O_3. 4SiO_2$	HIDDENITE	Monoclinic	6·5	3·18	1·66–1·67	G
$CaMg_3Si_4O_{12}$ or $CaO. 3MgO. 4SiO_2$	JADE (Nephrite)	Monoclinic	6·5	3·00	1·61–1·62	G, W &
$NaAlSi_2O_6$ or $Na_2O. Al_2O_3. 4SiO_2$	JADE (Jadeite)	Monoclinic	7·0	3·34	1·65–1·66	W, G, R, Br,
SiO_2 (and about 20% impurities)	JASPER	Crypto-Cry.	7·0	2·65	1·54	Br R Y &
$LiAl(SiO_3)_2$ or $Li_2O. Al_2O_3. 4SiO_2$	KUNZITE	Monoclinic	6·5	3·18	1·66–1·67	Pk
$NaCa_4Al_5Si_9O_{31}$ or $Na_2O. 6CaO. 7Al_2O_3. 18 SiO_2$	LABRADORITE	Triclinic	6·0	2·69	1·56–1·57	Gy-Iridescent
Very variable containing Silica (SiO_2); Alumina	LAPIS-LAZULI	Cubic	5·5	2·85	1·50	B & W
$LiNbO_3$	LITHIUM NIOBATE	Trigonal	5·50	4·64	2·21–2·30	W, G, R, Y

Refrac- tion	Diapha- neity	Principal Localities	Remarks
—	L-O	S. America, India, U.S.A., Scotland	Used for industrial and ornamental purposes.
Double	T-L	Ceylon, Urals, and Rhodesia.	Imitated by synthetic corundum and spinel.
—	O	N. and S. America, Urals, Africa, India, Madagascar.	Subject to damage through easy cleavage.
Double	T-L	Ceylon, Africa, S. America, Urals, U.S.A., India	In jewellery the most important violet stone.
Double	T-L	Brazil, Ceylon, Africa, Madagascar, Siberia, Urals, U.S.A.	The most prized to-day are the sky blue stones.
—	O	India, Urals.	A massive quartz with inclusions giving a schiller.
Double	T-L	India, S. Africa, S. America, Urals, U.S.A.	Emerald, aquamarine, morganite are beryls.
—	O	Australia, Brazil, India.	Green with red spots this stone is used as a seal stone.
—	L	India and China.	A jade simulant. Wrongly called "new jade".
Double	T	Scotland, U.S.A., S. America.	Brown quartz often used for Scottish jewellery.
—	L	India, S. America.	Brownish-red chalcedony is sometimes called carnelian.
Double	L-O	Ceylon.	Chrysoberyl cat's-eye is the more valuable and is heavier than quartz cat's-eyes.
Double	L-O	Ceylon, Europe, India.	The ray in quartz cat's eye is less sharp.
—	L-O	World wide.	There are many coloured varieties, including a chrome green.
Double	T-L	Ceylon, Urals, Brazil.	Yellow chrysoberyls were used in Victorian jewellery.
—	L-O	India, N. America, Urals, U.S.A., Australia	Some agate is stained so as to resemble chrysoprase.
Double	T	N. and S. America, Urals.	Much citrine is amethyst turned yellow by heating.
Double	T	Burma, Siam, Africa and India and U.S.A.	The species which provides rubies and sapphires, also star stones.
Double	T	Man made: Germany, Switzerland, France and U.S.A.	Made in all colours except emerald green. Also as star stones.
Single	T-O	Africa, Brazil, India, Russia, British Guiana, Venezuela.	The hardest substance known to nature.
—	O	Israel	Usually tumbled.
Double	T-L	Australia, Colombia, India, S. Africa, Urals, Pakistan, Rhodesia.	The most highly prized variety of beryl.
(Slightly)	T-L	Grown artificially in the U.S.A., Germany and France.	Generally has lower S.G., R.I. and D.R. than natural emerald.
Single	T-L	England, U.S.A. and S.W. Africa.	The massive variety is known as "Blue John".
Single	T-O	Australia, Ceylon, India, S. America, U.S.A., Madagascar, Africa.	When cut in the cabochon style are called carbuncles.
Single	O	Urals, Italy, Switzerland.	Sometimes used for mourning jewellery.
Single	T	Urals.	The most valuable of the garnets.
Single	T	Africa and Pakistan	A recent discovery.
Single	L-O	South Africa.	The green material has been miscalled "Transvaal jade".
Single	T	Ceylon, Piedmont, Sweden, Brazil, U.S.S.R., U.S.A.	Has characteristic orange-yellow colour.
Single	T	Czechoslovakia, S. Africa, U.S.A.	Used in much jewellery of the Victorian period.
Single	T	U.S.A., Africa, Ceylon	Rhododendron-red colour.
Single	T	U.S.A., Madagascar.	Rare and not often found in jewellery.
Single	L-O	—	A brown or blue glass with included copper crystals.
—	O	Brazil, England, India, Scotland, Elba.	Used for intaglios and as an imitation of black pearl.
Double	T	N. Carolina, Brazil.	A rare stone resembling emerald.
—	L-O	U.S.A., Canada, New Zealand, Siberia, Rhodesia.	Used for carvings, beads and other small articles.
—	L-O	Upper Burma, California, Japan.	The most prized of the jades. The so-called "Chinese jade".
—	O	Urals, N. and S. America and other sources.	A colourful massive material used for small objects.
Double	T	Brazil, U.S.A. and Madagascar.	This stone is the lilac-pink variety of spodumene.
Single	L-O	Labrador, Russia, U.S.A.	A rock-like mineral with a play of colours, also transparent yellow.
—	O	Afghanistan, Chile, Siberia, California.	A deep blue ornamental stone. Often contains pyrites.
Double	T	Man made.	Sold as "Linobate".

Chemical Formula	Name of Gem	System of Crystallography	Mohs's Hardness	Specific Gravity	Refractive Index	Colo
$CuCO_3. Cu(OH)_2$	MALACHITE	Monoclinic	3·5	3·80	—	G
$CaCO_3$	MARBLE	Trigonal	3·0	2·71	1·48–1·65	W &
FeS_2	MARCASITE (pyrites)	Cubic	6·0	5·00	—	Metall
SiO_2	MOCHA STONE	Crypto-Cry.	7·0	2·62	1·53	W with RG
$KAlSi_3O_8$ or $K_2O. Al_2O_3. 6SiO_2$	MOONSTONE	Monoclinic	6·0	2·55	1·53–1·54	W & B Cha
$Be_3Al_2Si_6O_{18}$ or $3BeO. Al_2O_3. 6SiO_2$	MORGANITE	Hexagonal	7·5	2·7–2·9	1·57–1·58	Pk
SiO_2	MOSS AGATE	Crypto-Cry.	7·0	2·62	1·53	W with RG
$CaMg_3Si_4O_{12}$ or $CaO. 3MgO. 4SiO_2$	NEPHRITE	Monoclinic	6·5	3·00	1·61–1·62	G &
SiO_2	ONYX	Crypto-Cry.	7·0	2·65	1·54–1·55	Bk & W
$3SiO_2. H_2O$ (varies)	OPAL	Amorphous	6·0	2·10	1·45	W Bk & R
Mg_2SiO_4 or $2MgO. SiO_2$	PERIDOT	Rhombic	6·5	3·33	1·65–1·69	Y-G
SiO_2	PRASE	Crypto-Cry.	7·0	2·62	1·53	G
$MnCO_3$	RHODOCHROSITE	Trigonal	4·0	3·60	1·60–1·82	Pk
$MnSiO_3$ or $MnO. SiO_2$	RHODONITE	Triclinic	6·0	3·54	1·72	Pk Fleck
SiO_2	ROCK CRYSTAL	Trigonal	7·0	2·65	1·54–1·55	W
SiO_2	ROSE QUARTZ	Trigonal	7·0	2·65	1·54–1·55	Pk
Al_2O_3	RUBY	Trigonal	9·0	3·99	1·76–1·77	R
TiO_2	RUTILE (Syn)	Tetragonal	6·0	4·25	2·62–2·90	W Y R
Al_2O_3	SAPPHIRE	Trigonal	9·0	3·99	1·76–1·77	B Y G W
SiO_2	SARD	Crypto-Cry.	7·0	2·62	1·53	R
SiO_2	SARDONYX	Crypto-Cry.	7·0	2·62	1·53	R & W B
$H_4Mg_3Si_2O_9$ or $3MgO. 2SiO_2. 2H_2O.$	SERPENTINE	Amorphous	3·0	2·60	1·57	G R
$H_2Mg_3(SiO_3)_4$	SOAPSTONE	Massive	2·0	2·7	1·54	W. Gy. B
$3NaAl.SiO_4./NaCl.$	SODALITE	Amorphous	5·5	2·30	1·48	B
$CaTiSiO_5$ or $CaO. TiO_2. SiO_2$	SPHENE	Monoclinic	5·5	3·40	1·90–2·00	Y Br &
$MgAl_2O_4$ or $MgO. Al_2O_3$	SPINEL	Cubic	8·0	3·60	1·72	R B G Y Bk
$MgO. 3·5Al_2O_3.$	SPINEL (Syn)	Cubic	8·0	3·63	1·73	W B G P
$LiAl(SiO_3)_2$ or $Li_2O. Al_2O_3 4SiO_2$	SPODUMENE	Monoclinic	6·5	3·18	1·66–1·67	Y Pk &
$SrTiO_3$	STRONTIUM TITANATE	Cubic	6·0	5·13	2·41	W
$CaNa_4Al_6Si_{11}O_{32}$ or $2CaO. 3NaO. 5Al_2O_3. 22SiO_2$	SUNSTONE	Triclinic	6·0	2·66	1·54–1·55	R-Br Spa
$Al_2(F,OH)_2SiO_4$	TOPAZ	Rhombic	8·0	3·53	1·63–1·64	W Y Pk B
$H_{11}Al_3B_2Si_4O_{21}$ (variable)	TOURMALINE	Trigonal	7·5	3·10	1·62–1·65	All colo
$Al_2(OH)_3PO_4./H_2O$	TURQUOISE	Amorphous	6·0	2·75	1·60	B
$Y_3Al_5O_{12}$	YTTRIUM ALUMINATE	Cubic	8·0	4·57	1·83	W &
Y_2O_3	YTTRIUM OXIDE	Cubic	8·0	4·84	1·92	W
$ZrSiO_4$ or $ZrO_2. SiO_2$	ZIRCON	Tetragonal	7·5	4·1–4·8	1·81–1·98	B W Y B
$Ca_2(Al, OH)Al_2(SiO_4)_3$	ZOISITE	Rhombic	6·0	3·35	1·69–1·70	B. M &
$C_{10}H_{16}O$	Organic Substances — AMBER	—	2·5	1·05–1·1	1·54	Y & R
$2MgCO_3. 21CaCO_3$	CORAL	. —	3·75	2·68	—	W Pk R
—	JET	Amorphous	3·5	1·33	1·66	Bk
––	PEARL	—	4·0	2·70	—	W Various T

KEY: Colour: B—Blue. Bk—Black. Br—Brown. G—Green. Gy—Grey. M—Mauve. Pk—Pink. R—Red. W—White. Y—Yellow. Diaphaneit

Refrac- tion	Diapha- neity	Principal Localities	Remarks
—	O	Australia, Urals, Congo and U.S.A.	A dark green banded ornamental stone.
—	L-O	Brazil, Argentina, Mexico and Algeria.	Also called "onyx marble".
—	O	England and most countries.	The mineral used for jewellery marcasite is pyrites.
Double	L	India, N. America.	May be said to be an alternative name for moss agate.
Double	T	Ceylon, N. America, Burma.	Moonstone when cut en-cabochon shows bluish gleams.
Double	T	U.S.A., Brazil.	The pink variety of beryl. Named after J. P. Morgan.
Double	L	China, India, N. America.	Agate with tree-like markings of red, green or black.
—	L-O	U.S.A., Canada, New Zealand, Siberia, Rhodesia.	The more common and less valued of the jades.
—	O	Czechoslovakia, Germany, India, S. America.	Often used for intaglios and cameos.
Single	T	Australia, Honduras, Hungary, Mexico, U.S.A.	Opal on a dark ground (Black opal) is the most prized. Some opal is stained.
Double	T	Red Sea, Burma, U.S.A.	Characterised by its oil green colour and strong D.R.
—	L-O	Finland, Germany, Scotland.	A dull green stone of the quartz group.
—	L-O	Argentine, U.S.A., and elsewhere.	A pink banded material used as an ornamental stone.
—	O	Russia, Sweden, U.S.A., Australia.	An ornamental stone pink in colour with black veins.
Double	T	World wide.	Has industrial uses as well as for beads and statuettes.
Double	L	U.S.A., Brazil, Ceylon, India, Urals.	Very rarely clear. Used for small carvings.
Double	T	Burma, Ceylon, India, Siam, U.S.A., Tanzania.	Some ruby when cut en-cabochon shows a six rayed star.
Double	T	Man made in America.	Stones have exceptional fire and double refraction.
Double	T	Australia, Burma, Ceylon, India, Siam, U.S.A., Africa	Sapphire, more commonly than ruby, produces starstones.
—	O	India, S. America.	A brownish-red chalcedony used for seal stones.
—	O	Germany, India, S. America.	Red and white banded chalcedony.
—	O	Canada, England, Ireland	A red and green ornamental stone, see also Bowenite.
Double	O	Canada, Africa, India	Used for carvings.
—	O	Canada, S.W. Africa.	Like lapis lazuli but a lighter blue and has pink spots.
Double	T	Alps, U.S.A., Mexico.	Sphene has more fire than diamond but is rather soft.
Single	T-O	Australia, Brazil, Burma, Ceylon, Siam, U.S.A.	Stones like ruby and sapphire but show no dichroism.
Single	T	Man made.	Made in many colours imitating stones of different species.
Double	T	Brazil, U.S.A., Madagascar.	The yellow stones are called spodumene (cf. kunzite)
Single	T	Man made in America.	A synthetic stone sold under the name "Fabulite" and "Diagem".
Double	L	N. America, Norway, Russia, Siberia.	Coloured schiller due to inclusions of an iron mineral.
Double	T-L	Australia, Ceylon, Madagascar, N. and S. America, Russia.	Values given for the brown and pink stones.
Double	T-O	Africa, Brazil, Burma, Ceylon, Urals, U.S.A.	Rubellite and indicolite are red and blue tourmalines.
—	O-semiO	Africa, U.S.A., Egypt, Madagascar, Persia.	American turquoise has lower density.
Single	T	Man made.	Sold as YAG, "Diamonair" and "Cirolite".
Single	T	Man made.	Diamond simulant.
Double	T	Australia, Ceylon, S. Africa, Indo-China, Burma.	B and W stones are heat treated from Indo-China rough.
Double	T	Tanzania.	Sold under name "Tanzanite".
—	L-O	Baltic, Burma, Sicily.	Sometimes contains insects and plant spores.
—	O	Waters of Japan, Mediterranean and Hawaii Islands.	Best colour is a deep rose red. Black coral has specific gravity 1·33.
—	O	England and Spain.	Jet is a type of hard coal.
—	O	Water of Australia, Indian Ocean, Japan, Mexico, Persian Gulf.	Freshwater pearls have less "orient" than oriental pearls. Pink conch pearls are non-nacreous.

parent. L—Translucent. O—Opaque.

*Crypto-Cry.—Crypto-Crystalline.
© 1972 N.A.G. Press Ltd., London.

Appendix VIII.
Comparison table of ring sizes

British Standard 1283:1945	European size	American size (approx.)	Inside diameter	
			inch	mm
A			.4750	12.065
	38		.4762	12.096
A½			.4828	12.263
			.486	12.344
	39	1	.4887	12.414
B			.4905	12.459
B½			.4983	12.657
	40		.5013	12.733
		1½	.502	12.751
C			.5060	12.852
C½	41		.5138	13.051
		2	.518	13.157
D			.5215	13.246
	42		.5263	13.369
D½			.5293	13.444
		2½	.534	13.564
E			.5370	13.640
	43		.5389	13.687
E½			.5448	13.838
		3	.550	13.970
	44		.5514	14.006
F			.5525	14.034
F½			.5603	14.232
	45		.5639	14.324
		3½	.566	14.376
G			.5680	14.427
G½			.5758	14.625
	46		.5765	14.642
		4	.582	14.783
H			.5835	14.821
	47		.5890	14.961
H½			.5913	15.019
		4½	.598	15.189
I			.5990	15.215
	48		.6015	15.279

British Standard 1283:1945	European size	American size (approx.)	Inside diameter	
			inch	mm
I½			.6068	15.413
	49	5	.6141	15.597
J			.6145	15.608
J½			.6223	15.806
	50		.6266	15.916
K		5½	.6300	16.002
K½			.6378	16.200
	51		.6391	16.234
L		6	.6455	16.396
	52		.6517	16.552
L½			.6533	16.594
M			.6610	16.789
		6½	.662	16.815
	53		.6642	16.871
M½			.6688	16.988
N			.6765	17.183
	54		.6767	17.189
		7	.678	17.221
N½			.6843	17.381
	55		.6893	17.507
O			.6920	17.577
		7½	.694	17.628
O½			.6998	17.775
	56		.7018	17.826
P			.7075	17.971
		8	.710	18.034
	57		.7143	18.144
P½			.7153	18.169
Q			.7230	18.364
		8½	.7265	18.453
	58		.7269	18.462
Q½			.7308	18.562
R			.7385	18.758
	59		.7394	18.780
		9	.743	18.872
R½			.7463	18.956
	60		.7519	19.098
S			.7540	19.152
		9½	.759	19.279
S½			.7618	19.350
	61		.7644	19.416
T			.7695	19.545
	62		.7770	19.736
T½			.7772	19.741
		10	.778	19.761
U			.7850	19.939
	63		.7895	20.054
U½			.7928	20.137
		10½	.794	20.168
V			.8005	20.333
	64		.8020	20.372
V½			.8083	20.531
		11	.811	20.599
	65		.8146	20.690

W			.8160	20.726
W½			.8238	20.925
	66	11½	.8271	21.008
X			.8315	21.120
X½			.8393	21.318
	67		.8396	21.326
		12	.843	21.412
Y			.8470	21.514
	68		.8522	21.645
Y½			.8548	21.712
		12½	.859	21.819
Z			.8625	21.908
	69		.8647	21.963
Z½			.8703	22.106
		13	.875	22.225

This table from BS 1283 : 1945, Jewellers' ring sticks and ring gauges is reproduced by permission of the British Standards Institution, 2 Park Street, London W1A 2BS, from whom copies of the complete standard may be obtained.

Note. Most BSI publications required for educational purposes may be bought by educational establishments at a discount of 40%.

Index